CHICAGO PUBLIC LIBRARY
WOODSON REGIONAL
9525 S HALSTED ST   60628

S0-CLC-594

```
HD          Fletcher, C. Edward.
4160.5
.F54        Privatization and the
1995        rebirth of capital
            markets in Hungary.
```

$36.50

DATE			

CHICAGO PUBLIC LIBRARY
WOODSON REGIONAL
9525 S. HALSTED ST   60628

BAKER & TAYLOR

*Privatization and the Rebirth
of Capital Markets in Hungary*

This work is dedicated to Mary,
who makes all things possible

Privatization and the Rebirth of Capital Markets in Hungary

by C. Edward Fletcher

McFarland & Company, Inc., Publishers
Jefferson, North Carolina, and London

British Library Cataloguing-in-Publication data are available

Library of Congress Cataloguing-in-Publication Data

Fletcher, C. Edward
    Privatization and the rebirth of capital markets in Hungary /
by C. Edward Fletcher.
      p. cm.
    Includes bibliographical references and index. ∞
    ISBN 0-89950-995-9 (lib. bdg. : 50# alk. paper)
    1. Privatization — Hungary. 2. Hungary — Economic policy —
1989–   I. Title.
HD4160.5.F54   1995
338.9439 — dc20   94-45043
                                                                                                           CIP

©1995 C. Edward Fletcher. All rights reserved

Manufactured in the United States of America

McFarland & Company, Inc., Publishers
  Box 611, Jefferson, North Carolina 28640

Acknowledgments

There are some people who assisted in the research or preparation of this book whose contributions were extraordinary. My outstanding research assistants, Lisa Falgiano, Elizabeth Rubin, and Deborah Spranger were particularly helpful. I am also deeply grateful to the many public officials and economic actors in Hungary who gave so generously of their time in educating me. Their names appear in the bibliography. Kriszta Bercz and Edit Jakab, my Hungarian language tutors, were more patient with me than I deserved.

Financial and logistical support came from a number of sources, including the International Research Exchange Board, the Council for the International Exchange of Scholars, the University of Cincinnati Center for Corporate Law, and the University of Cincinnati Joseph Fund for Corporate Law. I am grateful to all of them and to my dean, Joseph Tomain, who helped me secure funding and was patient enough to see the project through to completion.

But more than anyone else, I must thank my dear friends in Hungary who, without anything motivating them but the kindness of their hearts, set up interviews for me, made my life in Budapest possible, and helped me understand events in Hungary. More importantly, they taught me over the span of three years how deep and rich the human soul can be. With Tibor Madari, Zsuzsa Zatik, Zsolt Sztrokay, and Karoly Bognar in its citizenry, Hungary has cause for optimism.

Table of Contents

Acknowledgments   v

A Note to the Reader   xi

One: Introduction   1
    The Relevance and Unique Posture of Hungarian
       Capital Market Reforms   1
    The Role of Privatization in the Political and
       Economic Reforms of Hungary   8

**Two: The Legal and Institutional Framework for
Hungarian Privatization and Capital Market Reforms**   10
    Key Capital Reform Legislation: 1988–1992   10
       Act VI of 1988 on Business Organizations   11
       Act XXIV of 1988 on Investments of Foreigners in Hungary   14
       *Act XIII of 1989 on the Conversion of Economic
          Organizations and Business Associations*   15
       Act VI of 1990 on Securities and the Stock Exchange   16
       *Act VII of 1990 on the Foundation of the State Property
          Agency*   19
       *Act VIII of 1990 on the Protection of Property Entrusted
          to Enterprises of the State*   21
       *Act LXXIV of 1990 on the Privatization (Alienation,
          Utilization) of State-Owned Companies Engaged in Retail
          Trade, Catering, and Consumer Services*   21
       Act XVIII of 1991 on Accounting   22
       *Act XXV of 1991 on Partial Compensation for Damages
          Unlawfully Caused by the State to Properties Owned
          by Citizens in the Interest of Settling Ownership
          Relations*   23

Act IL of 1991 on Bankruptcy Procedures, Liquidation
  Procedures, and Final Settlement   24
Act LX of 1991 on the National Bank of Hungary   25
Act LXIII of 1991 on Investment Funds   25
Act LXIX of 1991 on Banks and Banking Activities   26
Supporting Institutions in the Reform Efforts: An Overview   26
  The Budapest Stock Exchange   27
  The State Securities Supervisor   27
  The Hungarian National Bank   28
  The State Property Agency   28

Three: The Sale Begins—The Origins of Privatization in Hungary 29

The State Property Agency: Origins, Organization,
  and Staffing   30
Why Privatize? And How Should It Be Done?   36
Setting Out: Pre-SPA Attempts at Privatization   42
  *Private Placements: The Examples of Tungsram and
    Graboplast*   42
  Pre-SPA Privatization: The IBUSZ Public Offering   44
  Spontaneous Privatization   47

Four: Privatization Under the State Property Agency— The Opening Programs 53

The First Privatization Program: Selling the Crown
  Jewels   54
The Second Privatization Program: Stripped Companies   63
The Third and Fourth Privatization Programs: Targeted
  Industries   65
The Small Retail Shop and Restaurant Privatization   66

Five: Changing Course in Privatization—More Programs with Greater Flexibility 70

Investor-Initiated Privatization   70
The Asset Management Program   72
Self-Privatization: Spontaneous Privatization Redux   74
Privatization Through Liquidation   78
The Breakup and Privatization of Hollow Parent
  Companies   80
The Asset Leasing Program   80

Table of Contents

Six: The Banking and Agricultural Sectors — Special Challenges in Reform and Privatization ... 82
  Reform of the Hungarian Banking Industry ... 82
    Banking System Reform, 1987–1992 ... 83
    Privatization of the Banking Sector ... 85
    The Ybl Bank Collapse and Other Scandals of 1992 ... 89
    The 1992 Bank Bailout ... 92
  Privatizing the Agricultural Sector ... 93

Seven: Another Mid-Course Correction — 1992 and the Shift Toward the Domestic Market ... 96
  Courting the Domestic Investor ... 96
    The Voucher Taboo ... 97
    But These Aren't Vouchers ... 99
      Preferential Loans for Domestic Purchasers ... 99
      Special Preferences for Workers and Managers of Privatized Firms ... 101
      Preference of Domestic Bidders Over Foreign Bidders ... 103
  The Status of Privatization Efforts at the End of 1992 ... 106

Eight: The Budapest Stock Exchange and the Securities Markets of Hungary ... 110
  The Old Budapest Commodity and Capital Exchange: 1864–1948 ... 111
  Trading Activities Prior to the Formation of the Budapest Stock Exchange ... 114
  Establishment and Organization of the New BSE ... 117
  Activities of the Exchange, On the Exchange and Off the Exchange ... 122

Nine: The Ongoing Debates — A Conceptual Framework for Thinking About Privatization Options ... 132
  How Fast to Go? ... 132
    The Pace Debate ... 132
    Measuring and Assessing the Slow Pace of Privatization ... 138
      Projections Not Met ... 138
      The Price of Slowness ... 139
      The Price of Speed ... 140
      Policy at the SPA Contributing to Pace Problems ... 141
  How Much Foreign Ownership Can Hungary Stand? ... 144
    The Problem Created by Perceptions ... 144

Reality: Some Data Concerning Foreign Ownership and Justifications for Significant Foreign Participation in the Process of Privatization	145
Should Shares Be Sold or Given Away?	148
Should Privatization Be Through Public Offerings or Private Placements?	150
Should the Process of Privatization Be Centralized or Decentralized?	152
How Should Shares Be Distributed Among the Many Claimants?	154
How Should the Proceeds of Privatization Be Spent?	156
How Much Residual State Ownership Is Desirable?	157

Ten: The Problems of Privatization—"The Process Is Far More Complicated Than We Ever Thought" 160

The Political Problem of Foreign Ownership	160
Lack of Demand	161
Domestic	161
Foreign	163
Lack of Credit	163
Valuation Difficulties	163
Overvaluation by the SPA	164
Political Distractions	165
Lack of Clarity of Property Ownership	166
Insufficient Financial and Legal Infrastructure	167
International Conflict	168
Bureaucracy, Controversy, and Confusion at the SPA	169
The Bad and Worsening Condition of Many State-Owned Enterprises	170
Lack of Qualified Staffing at the SPA	171
Possible Conflicts Among Laws	171
Telecommunications Legislation	171
Compensation Legislation	172
Environmental Problems	173

Eleven: What Do We Make of All This?—Some Lessons to Be Learned from the Hungarian Privatization Process 175

Notes 181

Bibliography 225

Index 235

A Note to the Reader

Before beginning, I must alert the reader to two conventions used throughout this book.

The first is that diacritical marks for Hungarian vowels are omitted in both Hungarian names and Hungarian words.

The second is the standardized conversion of Hungarian forints (HuF) to U.S. dollars at the rate of 75 forints per dollar. I have made that conversion at various points in the book either to give the reader some sense of the magnitude of the figure referred to or to provide a frame of reference for the comparison of figures being used. Wherever I have made that conversion, the precise figure, in dollars as converted at the exact exchange rate prevailing at the time in question, is unimportant. But the reader should understand that the exchange rate used throughout this book is an approximation. From 1988 to 1993, the value of the forint declined against the U.S. dollar, from approximately 65 to the dollar to approximately 100 to the dollar. Since most of the events described in the book took place in the period from 1988 to 1992, the 75-forint-per-dollar convention is pretty close to the average actual exchange rate.

One

Introduction

> *Forty years after Hungary took everything from the people, it's not easy to sell it back.*[1]
>
> *Just who is going to buy all the companies the ... Hungarians plan to sell?*[2]

The aim of this book is to describe and analyze the way in which the Republic of Hungary went about the process of privatizing its state assets, reforming its capital markets, and reforming the legal regimes that regulated those capital markets in the period 1982–1992. Thus this book represents a sort of modern history of certain economic and legal events in Hungary. Events in Central and Eastern Europe in the late 1980s and early 1990s have fascinated Western observers for a variety of reasons. But why study Hungarian privatization and capital market reforms?

The Relevance and Unique Posture of Hungarian Capital Market Reforms

In one sense the above question can be answered simply by saying that this modern history is important for the reasons that history is important generally. But there are two more specific reasons as well, and they both stem from the fact that the Hungarian experiment in capital market reform and regulation has been undertaken under unique circumstances. Indeed, that experiment is part of a larger, unprecedented, phenomenon: the transformation of highly developed command economies into highly developed market economies.[3]

In order to discuss Hungary's system of capital market regulation, we must first define some terms. The word *capital* should be understood for present purposes in the broad sense of means of production: factories, shops, equipment, machinery, vehicles, farmland, etc. The term *capital markets* refers to the trading of interests in that capital. In a highly developed economy, that may take the form of, among other things, one or more securities exchanges. In a less developed economy capital markets nearly always exist as well, just on a smaller and perhaps more rudimentary scale. Agricultural land is bought and sold; shops and other businesses are opened and closed, and interests in those businesses may be divided among family members or other allied interests. The term *capital market regulation* simply refers to the rules that represent that subset of the society's legal system affecting the transfer or trading of interests in the society's means of production. In a complex market economy, a complex system of securities regulation usually exists. In a less complex market economy, there may not be volumes and volumes of statutes and cases refining the details of the rules governing capital market transactions, but there are likely to be very basic rules concerning, for example, methods of transfer, the rights of buyers and sellers, and proscriptions of fraud. These rules also constitute a system of capital market regulation.

One can think of the components capital, capital markets, and capital market regulation in terms of a game metaphor. The capital markets are the game; the capital represents the pieces with which the game is played; and capital market regulation constitutes the rulebook for the game. Ordinarily, those three components develop together over time, organically, in a sort of dance. The process is a familiar one to those familiar with the history of capital market regulation in nearly any developed market economy. And, importantly, the process historically has always been the same (at least in its basics) in the various systems.

First there will exist only a small amount of capital, typically agricultural land and some rudimentary equipment. That capital will occasionally pass from person to person, either by sale, inheritance, or force, and sometimes will be split into smaller interests or combined to form larger interests. In other words, the society typically begins with a small amount of capital and a rudimentary capital market. Along side those two components, there will exist a simple system of capital market regulation. The governing apparatus for the society, regardless of form, will stipulate (whether formally or merely by custom) certain norms of behavior in the transfer of that capital.

As the economy of the society develops and becomes more complex, the quantity of capital will increase. So too will the variety. As the capital structure of the economy becomes more complex, the capital market becomes more complex. The more stuff there is for trading, the more trading will take place. The number, size, and complexity of capital transactions will increase. Concommitantly, the system of capital market regulation will become more complex. Variety and volume of capital market transactions require a more sophisticated set of rules governing those transactions.

In typical market economies, this process continues as the economy develops. Not surprisingly then, the complexity of a country's system of capital market regulation correlates roughly with the complexity of its economy and the complexity of its capital markets. Western economies are typical in this regard. As the amount of capital has increased over time, the capital markets have become more sophisticated, and the bodies of capital market regulation have become more complex. Thus, for example, the complex systems of securities regulation in the United States and the United Kingdom have developed organically over time to keep pace with ever more complex capital markets, which themselves have developed over time to account for the increasing quantity and complexity of capital.

But the new market economies of the former Socialist countries in Central and Eastern Europe and the former Soviet Union are different. These are relatively highly developed economies—economies with a lot of capital in complex forms. But because of the Socialist systems, that capital has not been traded privately since being put under Socialism, at least not to a degree that would correspond to the quantity and complexity of the capital in existence. As a result, there has developed no complex system of capital market regulation. In that respect they are different from developing countries with either authoritarian or democratic governments and also different from developed countries with democratic governments. Thus in Hungary the capital (in the form of plants and equipment and the like) has been there all along. But that capital has not been traded privately in any meaningful respect since shortly after the Second World War. And because of the retarded nature of the capital markets, Hungary did not have until recently a system of capital market regulation that would reflect the quantity and complexity of the capital available.

Today in Hungary and in other former Socialist countries, that is changing, but the process is not the typical organic development described above. Instead, the capital is being placed in private hands rapidly (relatively speaking), and capital markets have sprung up almost overnight as a result. That has required Hungary and other former Socialist countries to develop systems of capital market regulation quickly, without the luxury of allowing the system of securities regulation to develop organically over time as the quantity of capital increases and the capital markets develop. This makes the development of capital market regulation in former Socialist countries unique. The development has had to be more rapid and complete than it would have been had these economies developed in the way other market economies, such as those of the United States and the United Kingdom, have evolved.

This lack of organic development represents both a curse and a luxury to Hungary. The difficulty is that Hungary has been forced to create rules for a game it has little experience in, a game that will be played immediately at a relatively sophisticated level. It cannot move incrementally in its development of market and legal structures to the extent seen in the organic development of other market economies. As Plantagenet and Fiona Somerset Fry have written in the context of the development of a different legal system: "Legal systems generally take a long time to develop; yet there are safeguards in the slowness, for every step is constructed, tested, and amended where needed, and that in the long run is advantageous for everyone.[4]

But this comparative lack of incrementalism in Hungary has its benefits as well. Hungary can invent whatever structures and legal regimes it sees fit, without being tied to overly complex and outdated formulae whose existence results more from history than logic. Hungary is writing on a comparatively clean slate, and that is an opportunity. Or, to use a common metaphor of law,[5] one can see the process of legal development as akin to the development of barnacles on the hull of a ship. Barnacles build on the hull over time and eventually create a hazard to seaworthiness; they must be scraped off from time to time. So too with capital market regulation in developed economies: the systems become more and more complex over time as layer upon layer of regulation is added. Eventually that level of complexity is more a hindrance to the market than an aid to it. Hungary sails with a clean hull.

But we must not make too much of the idea that Hungary is, with its former Socialist cousins, unique. Other countries outside the Socialist block have experienced privatization in various degrees. The 1980s saw a receding of statism, to greater or lesser extents, in the United Kingdom,[6] France,[7] Austria,[8] South America,[9] Africa,[10] Asia,[11] and in the United States.[12] And of course privatization efforts continue in other parts of the world, most notably in India.[13] Still, none of those privatization programs has had the scope of Hungary's. Privatization in the United Kingdom, for example, resulted in transformation of only approximately 5 percent of the gross domestic product in the 1980s[14]; Hungary hopes to privatize more than 50 percent of its state-owned enterprises by 1994[15] and 50 percent of its farmland over approximately the same period.[16] The point was made clearest by Lajos Bokros, a director of Hungary's State Property Agency, which oversees privatization in that country: "It is the speediest privatization in history."[17]

And the problems attending privatization in one country are often different in important respects from those in others.[18] For example, David Bartlett has suggested that privatization efforts are more easily accomplished in developed economies with democratic institutional histories than in developing economies just emerging from autocracies.[19] We may expect Hungary, which has both a developed economy *and* a recent history of authoritarianism, to be different from either of those models.

It would also be a mistake to overdramatize the speed of privatization in Hungary and to ignore the degree to which Hungary is itself moving pragmatically, slowly, and organically in an attempt to avoid rushing headlong into systems that may prove ill advised and spin out of control.[20] Still, Hungary is unusual in meaningful ways in both the degree of change it has undertaken and in the pace of that change.

The unusualness of capital market regulatory development in former Socialist countries would by itself make the story of that development interesting and therefore worthy of study. But there are at least two more pragmatic reasons to study the process in Hungary.

The first is that Hungary is ahead of most former Socialist countries in its development of capital market legal structures.

Poland and Czechoslovakia (and later both the Czech Republic and Slovakia) have also moved quickly in this regard and in some respects (most notably in privatization and, in Poland at least, in currency regulation reform) are ahead of Hungary. Still, Hungary has moved more rapidly than even these other two in setting up complex legal structures and capital market exchange mechanisms. The other former Socialist countries lag far behind. But they too will probably go through a process similar to that in Hungary. As of the end of 1992, Russia, Ukraine, Estonia, Latvia, and Lithuania were already taking significant steps in that regard. The others, notably Croatia, Slovenia, Byelorus, Bulgaria, and Romania, will likely follow to varying degrees. As these other former Socialist countries replicate, in their own unique circumstances, the process of privatizing their economies, they too will be faced with the need to develop systems of capital markets and capital market regulation without the luxury or curse of organic development. As they do, it would be both natural and wise for them to learn what they could from countries such as Hungary that have gone before them. This fact is not lost on the Hungarians. As one observer noted early in the process:

> Former socialist countries, starting privatization later, will be in a more advantageous position, since they will probably be able to avoid numerous mistakes, learning from ours, thus being also able to adapt well-proven methods.[21]

Not surprisingly then, in the early 1990s, Russia was looking at Hungary as a possible model for its own capital market regulatory reform and sending delegations to Budapest to observe Hungarian capital market institutions.[22]

Because Hungary may serve as a model for future capital market reform and regulation, the process it has gone through and the lessons it has learned are worth describing. But highly developed market-economy nations have something to learn from the Hungarian experience as well. During the period of market and regulatory reform in Hungary, developed Western countries have too often assumed only the role of teacher vis-à-vis the Hungarians, with almost no attempt to assume the role of student. But there is much to learn. Because it is writing on a relatively clean slate, Hungary can attempt to achieve an optimal level of complexity in its system of securities regulation. What Hungary had come up with as of the end of 1992 was certainly less complex than corresponding systems in the United States, the United Kingdom, and other developed countries. In some aspects, Hungary has not provided enough complexity in its legal regimes to account for the complexity in the capital markets it has created. In other respects, the simplicity of its legal rules has worked just fine. By examining the Hungarian example, observers in more developed market countries can tell a lot about what is essential and what is inessential in a system of capital market regulation. The Hungarian example also offers us valuable lessons concerning the desirable degree of state ownership of the means of production in an economy. Of course just as we must not make too much of the uniqueness of the Hungarian example, we must be careful about assuming too much concerning the transferability of the lessons learned by the Hungarians. What works (or fails) in Hungary may not work (or fail) in another country, due to local circumstances.

History is of course seamless, and for that reason any study with temporal parameters is at best arbitrarily segmentary and runs the risk of being misleading in portraying the episode without reference to what came before and after the period under study. But it is the nature of a story that it have both a beginning and an ending. The trick is to find a logical place to start and finish. This book tells a story that begins in 1982 and ends in 1992, although it deals (briefly and to the extent necessary) with both earlier and later periods.

For example, as described in later chapters, Hungary had a highly developed stock exchange from the late nineteenth through the middle of the twentieth centuries. And Janos Kadar's New Economic Mechanism, begun in 1968, set the stage for many of the reforms that took place in the 1980s and 1990s. The year 1982 is also an odd place to begin because at that time Hungary was firmly in the Socialist sphere of influence, both politically and economically. Nonetheless, it is a good place to begin, because it was in 1982 that Hungarian companies were first permitted to sell bonds to the public as a means of generating capital.[23]

By ending with 1992, this book stops more in the middle of the story than at the conclusion. But one must stop somewhere, and by the end of 1992, the most important legal and economic structures had been put in place in Hungary, and it was just a matter of fine-tuning and waiting to see how everything would come out. At the end of 1992, the attitude of both Hungarian and foreign observers was generally one of cautious optimism.

But that is getting ahead of the story. Before beginning our examination of Hungarian privatization and the rebirth of the Hungarian capital markets, we must understand the central role that privatization played in the economic and political reforms in Hungary in the late 1980s and early 1990s.

The Role of Privatization in the Political and Economic Reforms of Hungary

In 1987 what we now know as the Republic of Hungary was still known as the People's Republic of Hungary. But in that year the ruling Hungarian Socialist Workers party, which had been running a Soviet-style Socialist state since 1948, began an ambitious set of reforms that would ultimately transform the legal and economic structure of the country in powerful ways. The state-owned economy was to be privatized, the banking system was to be reformed to work more like its Western counterparts, the Budapest Stock Exchange would be reopened after a hiatus of 42 years, a new regulatory agency (the State Securities Supervisor) would be given authority over the capital markets, and more than a dozen major pieces of legislation would transform the legal system surrounding Hungary's business environment in ways designed to facilitate a transformation to a private market economy.

Much of this book is about the legal and regulatory structures and processes of Hungarian privatization. But to understand those structures and processes, a study of the context of that law and regulation — the business and economics surrounding them — is necessary. In addition, privatization, banking reform, development of the Budapest Stock Exchange, enactment of reform legislation, and establishment of the State Securities Supervisor together form a web of interrelated reforms of the Hungarian capital market infrastructure, no one of which can be understood except in relation to the others. As one journalist wrote the matter in July 1990, "Privatization is just one part of a total economic reform package, albeit perhaps the most important part."[24]

The government has clearly understood since the early days of reform both the importance of privatization and its relation to other economic reforms. To achieve the goals of privatization, said a government economist in early 1990, "We must stabilize the forint, pay a real interest rate, and nurture private enterprise."[25]

International organizations have also seen the important role privatization of the Hungarian economy plays in forwarding and solidifying the Westernization of Hungary. In February 1992, the World Bank announced it would lend Hungary $200 million to finance privatization efforts, with a plan to lend approximately $300 million per year after that.[26] And the United States government established an investment fund to support Hungarian privatization and investment through the SEED Act (Support for East European Democracy),[27] a clear recognition of the interdependence of economic and political stability in the region.

The Hungarian privatization efforts rest on a legal and institutional framework with which one must be familiar at least on a basic level in order to understand those efforts. Chapter Two therefore describes the aspects of Hungary's legal framework and institutions that are most important to privatization.

Two

The Legal and Institutional Framework for Hungarian Privatization and Capital Market Reforms

From 1988 to 1992 the Hungarian Parliament[28] passed thirteen laws (together with amendments to those laws) that can be described as fundamental for purposes of privatization. In addition, two institutions—the Budapest Stock Exchange and the State Securities Supervisor—were established in 1990 and have played important (albeit unexpectedly peripheral) roles in privatization. This chapter examines first the laws, then the institutions. The role of the most important privatization institution, the State Property Agency, is described at length throughout this book and is therefore not taken up separately in this chapter.

Key Capital Reform Legislation: 1988-1992

The law in Hungary, like the law everywhere, is a seamless web. Most aspects of the Hungarian legal system impact privatization to one extent or another.[29] This section, however, presents thirteen statutes passed by the Hungarian Parliament in the late 1980s and early 1990s that are fundamental to privatization. Familiarity with these statutes is necessary for an understanding of the shape of privatization efforts described in subsequent parts of this article. For simplicity, these statutes are presented here in chronological order.

The Legal and Institutional Framework 11

Act VI of 1988 on Business Organizations[30]

The 1959 Civil Code of Hungary recognized business associations but treated them only cursorily as special contractual relationships.[31] It was not until the Act on Business Organizations was passed in 1988 that a variety of business associations was provided for and regulated in detail. There had been a company law operating before the ascendancy of the Hungarian Socialist Workers party, but it was more than 100 years old and of little use.[32] Still, according to one academic, certain ideas from the old statute were carried over into the new act, and that old statute might prove useful if there were an interpretive difficulty in the new act.[33]

The Act on Business Organizations was consciously based on modern Western European/American models. It contains three parts, only two of which are substantive. The first contains introductory provisions detailing, for example, the purpose of the legislation[34] and the scope of the legislation.[35] It also guarantees foreign interests the repatriation of capital and profit in the currency of the foreigner's contribution to the enterprise, notwithstanding the limited convertibility of the Hungarian forint.[36] Part I of the act also contains general company-law rules applicable to all business associations, regardless of type. For example, all business organizations are formed by the signing of articles of association, which must then be filed with the appropriate court of registry. The existence of the organization begins with the filing, but once the filing is made, that existence is retroactive to the date of signing.[37] The first part of the act also provides for supervision by supervisory boards, senior officers, and auditors; provides for judicial review of company resolutions; and stipulates rules for winding up an association.[38] Interestingly, Hungary has allowed a great deal of flexibility in the rights and duties of the various parties to all types of business organizations (except one) by providing that, except where expressly prohibited by law, any provisions of the act may be varied by agreement.[39] This freedom to vary legal rules by contract was added to the act in 1991.[40]

Part II of the statute provides separate rules for the six types of business organizations contemplated by the act. They are the general or unlimited partnership (kozkereseti tarsasag or "Kkt."), the limited partnership (beteti tarsasag or "Bt."), the business association (egyesules), the joint venture (kozos vallalat or "Kv."), the limited liability company (korlatolt felelossegu tarsasag or "Kft."), and the public company limited by shares (reszveny tarsasag or "Rt.").[41] Each has its Western counterpart, and each has its own rules governing its organization, capitalization, and operation.

The general partnership (Kkt.) is formed by the signing of articles of association that include an obligation to pursue, with unlimited joint liability of the partners, a common business activity.[42] The partners enjoy joint management rights,[43] although the articles of association may appoint one or more partners to perform this function.[44] Except for special matters requiring unanimity or a supermajority, partnership decisions are to be resolved by majority vote of the partners.[45] The articles of association may provide for supermajorities in matters not mandated by the act, but where the act mandates unanimity or a supermajority, the articles of association may not vary.[46] Partners are jointly and severally liable for all obligations of the partnership, even if those obligations were incurred before a given partner joined the partnership.[47] This joint and several liability may not be waived by contract[48] and continues for five years after a partner leaves the partnership with respect to obligations incurred while he or she was a partner.[49]

The limited partnership (Bt.) is formed by the signing of articles of association, for conducting a business activity, that contain the following provisions: (1) At least one active partner must bear unlimited liability. If there is more than one active partner, that unlimited liability is joint. (2) There must be at least one passive partner whose liability is limited to his capital contribution.[50] Only the active partners are obligated and entitled to manage the partnership,[51] and only the active partners may represent the partnership in dealings with third parties.[52] These provisions may not be varied by agreement.[53]

The business association (egyesules) is less a business organization than a trade association. It acts on a nonprofit basis to further the joint interests of other economic associations.[54] Its members are jointly and severally liable for its obligations.[55]

The joint venture (Kv.) is an association of business entities operating for profit in which the respective profits and liabilities are allocated among members according to their capital contributions,[56] although these allocations may be modified in the articles of association.[57] Control of the joint venture is in the hands of the board of directors,[58] which must meet at least once a year.[59]

Management is exercised by a managing director[60] (or directorate if the number of venture members so justifies[61]), who operates under the control of the board of directors.[62]

The limited liability company (Kft.) is modeled on the German GmbH (Gesellschaft mit beschrankter Haftung) and is similar to the American closely held corporation. Its distinguishing characteristics are the limitation on shareholder liability to the amount of shareholder capital contribution, the prohibition on public solicitation for shareholders, the requirement that each shareholder also provide services of tangible value to the company, and the requirement that initial capital be at least 1 million *HuF* (approximately $13,000).[63] Thirty percent of the capital must be in cash, and no shareholder may contribute less than 100,000 *HuF* ($1300). A Kft. may be formed as a one-person company.[64] Shares may be transferred to persons not previously shareholders only after offering them first to the company.[65] The management and control of the limited liability company are in the hands of the meeting of members (i.e., shareholders' meeting),[66] the managing director,[67] and the supervisory board.[68]

The public company limited by shares (Rt.) is modeled on the German AG (Aktiengesellschaft) and is similar to publicly held corporations in the United States. As in the Kft., the shareholders' liability in the Rt. is limited to their capital contribution.[69] Unlike the Kft., however, variations by agreement from the legal rules governing the Rt. are only possible where the act specifically so provides,[70] minimum initial capital is 10 million *HuF* (approximately $130,000),[71] and there is no restriction on public solicitation of shareholders or a requirement that shareholders work for the company. As in the limited liability company, company governance is in the hands of the company meeting (i.e., shareholders' meeting),[72] a managing directorate,[73] and a supervisory board.[74]

Enactment of the Act on Business Organizations furthered privatization not only by providing potential foreign bidders with company structures with which they were familiar, but also by providing a shareholding form of company into which state owned firms could be transformed. Once a state enterprise was transformed into a limited liability company or a public company limited by shares, the bidder could make an acquisition simply by buying shares. And those shares entailed rights, defined by the act, that were similar in substance to those with which the bidder was already familiar. The act also provided both domestic and foreign bidders with a choice of forms in which to conduct business with former state assets, each with defined benefits and detriments.

Act XXIV of 1988 on Investments of Foreigners in Hungary[75]

Although the Act on Business Organizations provided explicit protection to foreigners buying or forming Hungarian businesses, it also explicitly provided that more detailed protections would come in a separate statute.[76] That separate statute, the Act on Foreign Investments, came only a few months after the Act on Business Organizations was passed.

Although both Hungary and the international community made much of the guarantee in the Act on Foreign Investments against nationalization,[77] the significance of such a statutory guarantee is unclear given that what a parliament enacts, it can abrogate. Much more important were the permission to acquire stakes in Hungarian companies and the substantial tax incentives included in the statute.

As originally enacted in 1988, the Act on Foreign Investments prohibited a foreign firm from acquiring a majority stake in a Hungarian firm.[78] It also required a firm established exclusively or primarily with foreign capital to obtain a license from the Ministry of Finance.[79] Both of those restrictions were lifted when the statute was amended later in 1990.[80] The statute as originaly drafted was still significant in that it explicitly permitted the establishment of new wholly foreign-owned firms and the purchase by foreigners of minority stakes in Hungarian firms.

The tax incentives to foreign investors were substantial. If a company with registered capital of at least 50 million HUF and foreign investment of at least 50 percent garners at least half its revenue from the manufacture of goods or operation of a hotel erected by the company, it is entitled to a 60 percent tax break for the first five years and a 40 percent tax break for the second five years.[81] In addition, if more than half the revenue of the company comes from certain "particularly important activities" listed in an appendix to the statute, the tax break is 100 percent for the first five years and 60 percent for the second five years.[82] In addition, the firm is entitled to a tax deduction to the extent its profits are reinvested in the firm,[83] and importation of necessary equipment is duty free.[84]

The effect of first permitting foreign acquisitions of Hungarian firms and then providing tax breaks to encourage these acquisitions obviously encouraged foreign participation in the privatization process. That it also put domestic bidders for state assets at a comparative disadvantage illustrates how important the encouragement of foreign investment in Hungary was in the early stages of privatization. As we will see below, foreign domination of the privatization process was later cited by the government as justification for attempts to bolster domestic demand for privatized firms.[85]

Act XIII of 1989 on the Conversion of Economic Organizations and Business Associations[86]

Having recognized the value to privatization of transforming state owned enterprises into share companies or other business organizations provided for in the Act on Business Organizations, Hungary enacted a statute to provide a framework for those transformations. The Transformation Act governs the conversion into business organizations of all types of state businesses, including cooperatives.[87] The act provides for a state business to begin the process by preparing a plan of transformation, which must include the business objective to be achieved by tranforming and the articles of association for the new form.[88] Whatever organ was entitled to make the decision to transform the firm must then publish the decision to transform and notify all creditors.[89] In the case of cooperatives, the management makes that decision. In cases of firms controlled by worker councils, those councils decide upon transformation. In the case of most other enterprises, the founding ministry or other state organ makes that decision.[90] Final approval must come from the State Property Agency.[91] Included in the statute are special rules for the various types of transformations contemplated (e.g., into a limited liability company or into a public company limited by shares).

When the firm is transformed, all shares not sold in the course of transformation to outside investors belong to the State Property Agency,[92] which also must approve of any sales to outside investors.[93] Once transformed, the new business organization becomes the full legal successor of the old state enterprise. As such, it succeeds to all legal rights of the old enterprise and inherits all liabilities as well.[94]

The one glaring omission in the Transformation Act is any positive requirement that state enterprises transform themselves into share companies that could then be privatized. Instead the act leaves much of the decision making to those in charge of the state enterprises, who may rightly see that transformation and privatization could adversely affect their own self-interest. As explored below, the privatization process took a disappointing pace, and part of the problem was the delays in transformation.[95]

Act VI of 1990 on Securities and the Stock Exchange[96]

Hungarian policy makers recognized early that privatization efforts would be aided by an efficient and properly regulated market for shares, both in initial offerings and in the secondary markets. That was one of the reasons for the enactment in January 1990 of the Hungarian Securities Act.

Securities, in the form of bonds, had been traded in Hungary since 1982, when the state began issuing government debt securities that could be traded. In 1988 the banks began meeting regularly to trade bonds.[97] Until the Securities Act was passed in 1990, however, the purchase and sale of securities was governed by the general provisions of the Hungarian Civil Code.

The Securities Act is brief and, in places, sketchy. The government clearly understood that the act was to be an outline and framework for future regulation rather than a comprehensive body of regulation that would forever serve the needs of Hungarian capital markets:

> With a view to the fact that the securities market is now in its infancy in Hungary, the present Act does not set forth detailed rules and regulations with respect to a number of issues; it merely determines the framework for such rules. ... The introduction of such a strict ... set of regulations would presumably not facilitate the development of this embryionic market; it would rather slow its progress. Therefore, efforts were made to arrive at compromises when elaborating the Act which would provide sufficient market development [and simultaneously be] good enough to ensure protection of investors.[98]

Still, some participants in the capital markets of Hungary found even this limited structure unnecessarily complex.[99]

The act itself is broken down into seven substantive parts and a set of closing provisions. Part I contains general provisions, including a statement of the scope of the act. Significantly, the act only applies to public issues of securities, defined as offers for sale to an unpredetermined group of possible investors.[100] Neither the number nor the sophistication of the offerees is deemed relevant. Rather, the only question is whether the offerees are specified or the offer is made to the general public. According to one of the authors of the statute, it would therefore be possible to make a placement to 2000 unsophisticated offerees and, so long as the offerees were named at least by category, the offering would not be regulated at all by the statute.[101] Indeed, according to a lawyer with a large Austrian investment bank, an offering limited to all persons listed in the Budapest telephone directory would be a private placement not covered by the act.[102] This led to an odd confrontation between the State Securities Supervisor and the State Property Agency. The SPA had been soliciting bids in newspapers for companies it was privatizing. These solicitations were aimed primarily at foreign investors who could bid for the entirety of the stake being privatized by the SPA. Although the advertisements were clearly directed toward potential private placements, they were not directed to any particular class of offerees. The SSS therefore took the position that they were "public offers" necessitating the use of a prospectus. Following negotiations between the two agencies, the SSS dropped its insistence on prospectuses in exchange for agreement by the SPA to ensure that the advertisements would not lead to widespread distributions. Thus, by interpretation, the SSS was moving toward a functional approach to the concept of "public offer."

Part II of the statute describes the role of the State Securities Supervisor.[103] Any public issuance of securities must be by means of a prospectus meeting the requirement of the act and approved by the SSS.[104] According to the head of the SSS, the agency has taken the position that it has unlimited power to insist that material be added or deleted from a prospectus and can refuse to permit an issuance if the prospectus does not comply with SSS requests.[105] The SSS is also given in Part II of the act the task of licensing brokerage firms and their employees, overseeing the activities of the Budapest Stock Exchange, and overseeing the brokerage industry generally.[106]

Part III of the act describes the method that must be used when an issuer wants to make a public offering of securities.[107] It also declares an issue "null and void" if securities are issued without an approved prospectus, although it does not state the legal effect of that declaration.[108] Finally, Part III requires certain periodic reporting by issuers.[109] Part IV of the statute concerns the retail brokerage industry but is primarily limited to the role of that industry in the public offerings that are regulated by the act.[110] The more general rules of professional conduct for securities brokers are contained in a separate statute on unfair business practices.[111]

The Budapest Stock Exchange is the subject of Part V of the statute. That part provides substantially for self-regulation on the part of the BSE but mandates the governance structure of the exchange.[112] It also stipulates the qualifications for membership on the exchange as well as grounds for termination of that membership and limits the types of business in which the exchange may engage.[113] It also provides for arbitration of disputes between exchange members, although not for arbitration of disputes between customers and exchange members.[114]

Part VI of the Securities Act contains detailed insider-dealing prohibitions. It was drafted intentionally to comply with the European Community's Insider-Dealing Directive. As such, it contains a specific prohibition on insider dealing together with definitions of "inside information," "insider," and "insider dealing."[115] In some circumstances trading by an insider is presumed to be insider dealing, although there are also provisions for affirmative defenses and an explicit statement of transactions that do not count as insider dealing.[116] According to an American lawyer with substantial experience in the Hungarian capital markets, if the Hungarian authorities seriously clamped down on insider dealing, business in Budapest would change substantially.[117] "Everything is done on inside information," he stated.[118] Without sophisticated monitoring mechanisms, insider dealing would be almost impossible to prevent.

Part VII of the statute concerns protection of investors and provides for recision by the purchasers in a public offering if the offering is not made pursuant to an approved prospectus.[119] In addition, it provides that the issuer and the issuing broker (i.e., the underwriter) are jointly and severally liable to the purchaser for any false information disseminated by them in the invitation to purchase.[120] But purchasers are not permitted private rights of action; rather, the public prosecutor or the sss may bring suit on behalf of the investors, and the outcome of that suit is binding on all investors.[121]

As the Hungarian capital markets become more sophisticated, the Securities Act will have to be amended to make it more than the mere outline it was intended to be. But one pervasive problem, mentioned by many involved with the act, is created by its generality and occasional ambiguity.[122] That problem involves the question of who is entitled to interpret the statute. Hungary has a long history of bureaucratic control. Much law has been created by ministerial decree, which makes it difficult to determine what the law is and undermines legitimacy. To overcome that problem, Hungary after 1990 attempted to ensure that as much law as possible came from the Hungarian Parliament. At the same time, however, Parliament could not attempt to deal with all the interpretive questions that might arise. Without specific rule-making authority, agencies charged with applying the statutes were sometimes left in an unclear position with respect to interpretation.[123]

At least for the first few years of its existence, the sss took the position that it had plenary authority to fill in the many gaps left by the statutory framework of the Securities Act.[124] This made application of the act speedy and efficient, but it resulted in a partial return to the regime of law by decree. It also made compliance with the law nearly impossible unless one were familiar with the various unpublished positions taken by the sss on various aspects of the statute.

Although the Securities Act is subject to criticism on a number of specific grounds and was even at its passage destined for revision, it represented an important step forward for the legal infrastructure that would support Hungary's capital markets generally and its privatization efforts specifically. It provided important guarantees to both domestic and foreign investors in those privatizations that took place by public offering. Perhaps more important, however, was the act's symbolic statement to foreign investors that Hungary was aligning its economic legislation with that of the West.

Act VII of 1990 on the Foundation of the State Property Agency[125]

Given the perceived importance of a well-regulated securities market to privatization efforts, it was perhaps no coincidence that immediately after passing the Securities Act, the Hungarian Parliament passed the Privatization Act. The act is divided into two substantive parts, one structural, the other procedural.

Part I creates the State Property Agency and describes its legal status, organizational structure, and tasks. It is a publicly financed institution not dependent upon revenue generated by it and is subject like other state agencies to oversight by Parliament and the State Audit Office.[126] Within the SPA, oversight is performed by an eleven-person board of directors. Five are to be representatives from five groups, including trade unions, employers' groups, and environmental groups.[127] The other six are appointed by Parliament for five-year terms.[128] Management is in the hands of a managing director, who is appointed by Parliament to a five-year term, but subject to dismissal by Parliament at any time.[129]

Pursuant to the act, substantially all state-owned productive assets were transferred to the ownership of the SPA, with exceptions for property owned by other state agencies and certain assets in the control of workers' groups.[130] The tasks of the SPA are detailed in the statute and fall into two broad categories: acting as portfolio manager for state assets and privatizing state assets.[131]

Part II of the act contains detailed rules for the privatization of state assets. Included is a provision permitting (but not requiring) assets to be sold by competitive tender.[132] In cases of substantially similar bids, the SPA is directed to prefer current management and employees to outside bidders.[133] The SPA was also authorized by the statute to lease assets rather than sell them.[134] Each of these three provisions would become important later. The SPA erred in practice too much on the side of caution and put all assets up for competitive bids even though it was not required to do so, thus substantially slowing down the process.[135] The requirement that the SPA prefer existing managers and employees reflected the ambivalence of the government toward what would surely be foreign domination of the privatization bids.[136] The SPA also took full advantage of its authority to lease assets, using that authority as one mechanism to devise creative methods of putting productive assets into private management.[137]

Although the SPA operated under some constraints imposed by the Privatization Act, the act also gave the SPA significant latitude to fulfill its tasks as it saw fit. Despite its brevity, the Privatization Act must be seen as the cornerstone of Hungary's privatization program.

Act VIII of 1990 on the Protection of Property Entrusted to Enterprises of the State[138]

As described below, the SPA and the Privatization Act came about as a result of scandals surrounding so-called "spontaneous privatizations" that took place without government control from 1988 to 1990.[139] These early privatizations were completed by managers of state-owned enterprises, and there was a widespread perception that those managers were not fulfilling their fiduciary duties to the state but were instead using privatization as a mechanism for self-enrichment. Thus, on the same day that the Parliament passed the Privatization Act, it also passed the State Property Protection Act.

The act was very short — only eleven sections in all — but it guaranteed state control of future privatizations by requiring all substantial sales or leases of state assets to be approved by the SPA.[140] "Substantial" sales or leases included several specific types of transactions, but in general they comprised those involving sales or leases of more than approximately $400,000 in value or 50 percent of the company, whichever was less.[141]

Act LXXIV of 1990 on the Privatization (Alienation, Utilization) of State-Owned Companies Engaged in Retail Trade, Catering, and Consumer Services[142]

The Hungarian Parliament also passed in 1990 a special statute for the privatization of small shops and restaurants. This program, whereby the approximately 8000 to 10,000 state-owned small retail and food establishments would be auctioned to domestic investors, is discussed at length below.[143] The statute authorizing the program is interesting primarily for the way in which the Parliament saw fit to provide a special statutorily prescribed program for this type of privatization rather than leave the privatization of those shops and restaurants to normal SPA operations.

The act required the SPA to initiate within two years a program to privatize those retail and catering businesses that averaged no more than 10 employees (and consumer services businesses averaging no more than 15 employees). The averages were based upon the period from December 31, 1988, until September 18, 1990, when the act became effective.[144] The businesses could be sold only to domestic natural persons or limited liability companies (Kft.'s) whose shareholders were domestic natural persons; foreigners were not eligible.[145] The statute called for auctions of the businesses, with the sales going to the highest bidders.[146] In cases of equal bids, priority was to be given to existing managers, existing employees, and persons paying in cash.[147]

As will become clear below, the program for these auctions failed miserably for a number of reasons. But the fact that the Parliament passed this statute is evidence that the government saw small business in a different light from large businesses and also that efforts were made early to ensure domestic participation in the creation of a private market economy.

Act XVIII of 1991 on Accounting[148]

The Hungarian Accounting Act is not remarkable because of its substance but because its passage was necessary. The act is a comprehensive (94 sections) set of basic Western-style accounting and bookkeeping principles that had not been part of Hungarian business. In mid-1990, for example, one Western accountant lamented the lack in Hungary of the concept of the nonperforming loan.[149] The importance for privatization of aligning Hungarian rules with those of the international community is made clear by the preamble to the Accounting Act, which states the purposes of the statute:

> In a market economy it is essential that the participants of the market — in order to be able to carefully prepare their decisions — have access to unbiased information. This information is primarily derived from previous records concerning the net assets and financial positions of ... organizations.
> The purpose of this Law is to provide accounting regulations — in line with international accounting principles — so that information can be provided which ensures a realistic and fair view of the income generating capabilities, net assets, financial positions and future strategies of those covered by this Law.[150]

As discussed below,[151] the process of privatization was slowed dramatically by the difficulty in valuing the companies being privatized. Not only was the future becoming increasingly difficult to predict, but the past performace of the companies (at least as represented on their books) was nearly incomprehensible to potential bidders. In fact, basic bookkeeping and accounting concepts were so foreign to Hungarian managers that seven months after the Accounting Act was passed, the government issued an official Explanation to Law XVIII of 1991 on Accounting[152] that reads like an introductory university text on accounting and bookkeeping principles.

Passage of the Accounting Act would not have an immediate effect on privatization, since its prospective nature ensured that the past performance of existing firms would be detailed only under the old system. But its passage made possible the facilitating of faster valuations for companies privatized in the latter half of the 1990s. In addition, it guaranteed some meaning to current statements of financial condition created after its passage.

Act XXV of 1991 on Partial Compensation for Damages Unlawfully Caused by the State to Properties Owned by Citizens in the Interest of Settling Ownership Relations[153]

One of the most fundamental legal questions hanging over Hungary's privatization efforts was that of compensation. If Hungary was going to put its state assets into private hands, what legal claims would be recognized for those from whom the state had taken the property some forty years earlier? Many argued that since the property had been taken illegally and without compensation, it should be returned to the people and their descendants at no charge. When Hungary's Constitutional Court ruled in 1990, however, that the government was not obligated to follow that path,[154] a legislative solution was clearly called for. That solution was the Compensation Act, passed by Parliament on June 26, 1991.

The basic idea of the statute was that the government would issue interest-bearing scrip, called compensation coupons, to those Hungarians (including émigrés) who could demonstrate that property had been taken from them illegally by the state after 1949.[155] The compensation coupons could then be used to purchase shares of privatized companies being offered to the public. And in fact, many shares were offered for compensation coupons. Typically, in a large public offer privatization, the SPA would set aside a percentage of shares to be offered in exchange for compensation coupons.[156]

In addition to the issuance of compensation coupons, the act contained a separate set of provisions for compensating former owners of farmland, whereby those former owners could be restored to the land taken from them or to nearby land.[157]

The number of claimants for compensation coupons and for farmland was a fraction of those expected by the government, perhaps in part because of the complexity of the process, the difficulty of producing the required documentation, and the extreme limits placed on the amount for which one could be compensated.[158]

And the effects of the Compensation Act's passage on privatization are difficult to measure. On the one hand, the act dispelled fears that previous owners would be found to have senior claims on property that had been taken. On the other hand, the act gave claimants the right to recover specific real property in certain limited circumstances, necessitating caution on the part of privatization participants bidding for property that could be subject to those claims. And only time will tell whether distribution of privatization shares for compensation coupons will create a broadly based shareholder class within Hungary. One thing is clear, however. By permitting domestic Hungarians to buy shares in privatized companies with compensation coupons, the government took a giant stride toward creating at least the impression that domestic Hungarians were participating in privatization in some meaningful way.

Act IL of 1991 on Bankruptcy Procedures, Liquidation Procedures, and Final Settlement[159]

The new Bankruptcy Act, passed by the Parliament on September 24, 1991, replaced a number of legal provisions that predated the market-oriented reforms of the late 1980s and early 1990s.[160] It was more evolutionary than revolutionary, expanding on the earlier principles to take account of a much larger private economy and bringing Hungarian law in line with other Western countries. Although the specifics of the act are unimportant for present purposes, the general importance of bankruptcy and liquidation for the privatization process should not be underestimated. As the moribund state enterprises sank lower and lower, liquidation became an important mechanism for privatization. Purchasers often found that state assets could be acquired more cheaply and quickly under the Bankruptcy Act than through the SPA. This subject is taken up below.[161]

Act LX of 1991 on the National Bank of Hungary[162]

For purposes of privatization, the two most important aspects of the Hungarian National Bank Act were its provisions putting the HNB in charge of monetary policy and interest rates[163] and its provisions permitting foreigners to acquire interests in formerly state-owned banks and financial institutions. As we will see, tight monetary policy and the concomitant unavailability of credit on reasonable terms together made it difficult for domestic Hungarians to purchase state assets, both in the sale of small shops and restaurants and in the sale of larger state enterprises.[164] In addition, the provisions authorizing foreign ownership of Hungarian banks were important in Hungary's attempts to overcome the special difficulties attending privatization of the banking industry. Those special difficulties are taken up below.[165]

There was also included in the act a provision that would require the Hungarian National Bank to divest itself entirely of any commercial bank shares it held within two years of the act's effective date.[166] This was designed to further both the decentralization of the commercial banking function away from the HNB and the privatization of the commercial banks.

Act LXIII of 1991 on Investment Funds[167]

As we will see, for a number of reasons very few privatizations took place by means of public offerings of securities to small investors. Rather, the vast majority were through private placements to lone large investors or small groups of large investors. As a result, domestic Hungarian individuals, most of whom had not accumulated large concentrations of investment capital, were effectively shut out of most of the transactions. They needed either credit or some way to pool their small investments.

The Investment Fund Act was designed in part to serve the latter need by permitting the creation of investment funds into which domestic small investors could put their limited savings, funds which could in turn invest that pooled money in privatized firms as well as provide needed equity capital to start- ups.[168]

The act authorizes both closed-end funds and open-end funds[169] and provides specific requirements for the fiduciary management of both types.[170] Included, for example, are rules requiring that capital reserves be set aside in open-end funds[171] and rules stipulating which types of investments are appropriate for fund managers to make.[172] Only persons licensed by the State Securities Supervisor are eligible to act as fund managers.[173] There are also special rules for the establishment and management of real estate funds.[174] The rights of investors, including both pecuniary and informational rights, are included in a separate set of provisions.[175]

The Investment Fund Act was passed by the Parliament in November 1991 and did not become effective until January 1, 1992.[176] For that reason, it was too early at the end of 1992 to measure the impact of the act on privatization. At that time, there were several funds being set up, and one (Creditanstalt's closed-end CA Investment Fund) was traded on the Budapest Stock Exchange. Passage of the act was an important attempt to facilitate the participation of more small domestic investors in the privatization process.

Act LXIX of 1991 on Banks and Banking Activities[177]

Most of the new Banking Act passed on November 13, 1991, consisted of much-needed tightening of the regulatory structure for the Hungarian banks, and most of it was only tangentially related to privatization. Some provisions, however, greatly affected how the banks themselves would be privatized. First, the act forbids any entity other than another banking institution from owning more than 25 percent of any Hungarian bank after January 1, 1997.[178] That means that the state's substantial (often majority) interest in the Hungarian commercial banks will have to be privatized. The statute also reduces voting rights for the state to no more than 25 percent as of January 1, 1995,[179] which has the effect of privatizing control even before privatization of ownership. Finally, the new statute requires as a practical matter foreign investment capital at most banks by requiring them to raise their capital levels and risk reserves to new minimums.[180]

Supporting Institutions in the Reform Efforts: An Overview

Together, the Budapest Stock Exchange, the State Securities Supervisor, the Hungarian National Bank, and the State Property

Agency comprised the primary elements of the institutional infrastructure for privatization. The first is taken up in Chapter Eight, the third is mentioned at various points in the book, and the last is the focus of many chapters that follow. Still, before beginning the story of privatization in Hungary, a brief overview of each is in order.

The Budapest Stock Exchange

The Budapest Stock Exchange's business is regulated at four levels: it is governed by the provisions of the Securities Act, it is subject to oversight by the State Securities Supervisor, it has its own charter, and its business is subject to its bylaws. For privatized firms' shares to be offered on the exchange, the SPA must satisfy the requirements of each level. Because the exchange offering would be a public offering, the SPA would have to satisfy the prospectus requirements of the Securities Act and gain the approval of the sss for the prospectus. The Council of the Budapest Stock Exchange would also have to approve the application of the SPA to have the shares traded there.

The State Securities Supervisor[181]

Like the Budapest Stock Exchange, the State Securities Supervisor is governed substantially by the terms of the Securities Act.[182] The agency was headed from its inception through 1992 by Dr. Zoltan Pacsi, who was appointed by the prime minister upon the advice of the finance minister. In January 1991 the sss had a total staff of ten, including one attorney, one secretary, one computer operator, and a few economists.[183] By late 1992, the staff numbered twenty-eight, and the entire agency had moved into new offices to accomodate their growing numbers.[184]

For purposes of privatization, the importance of the sss was at the end of 1992 mostly speculative. It has jurisdiction only over market participants (e.g., brokers and the BSE) and transactions involving public issues. Because only a handful of privatizations have been thought public issues, most sss work was unrelated to privatization in the first few years. Where it was involved, its role was limited to reviewing and approving the prospectuses being used. So long as the Securities Act continues to exclude from its coverage private placements, and until the SPA decides to privatize more companies through the public offering mechanism, the role of the SSS will continue to be limited in privatizations. Most observers, however, predict a greater emphasis by the SPA on the public offering, and the role of the SSS will grow accordingly. In addition, the SSS has always been an important regulatory link for public sales of securities, and the efficiency with which it works will partly determine whether the SPA decides to issue securities by means of a public offering.

The Hungarian National Bank

Until 1987, the importance of the Hungarian National Bank to the capital markets was twofold: it served as both the central bank of Hungary and the commercial bank for Hungarian businesses. In 1987 the commercial banking functions were spun off into several commercial banks, but the HNB retained the traditional central banking functions. Those functions were clarified and codified by two new and comprehensive banking laws enacted in 1991.[185]

Although very little of the history of privatization and capital market reform in Hungary focuses directly on the Hungarian National Bank, the presence and policies of the HNB underlie nearly all of it. For example, the HNB's credit policies directly affected the privatization efforts of the State Property Agency. In addition, most of the initial capitalization of the Budapest Stock Exchange came from the HNB. These relationships will become clearer in later chapters.

The State Property Agency

As the coordinating organ for all privatization efforts from 1990 onward, the State Property Agency is the central actor in the history of Hungary's efforts to transform its state economy into a private economy. A series of three laws passed by Parliament in 1990 put control of nearly all state-owned assets in the hands of the SPA and charged the SPA with the mammoth task of disposing of them.[186] The next three chapters detail the origins and activities of the SPA from 1990 to 1992.

Although much of this book takes up the various aspects of privatization in Hungary topically, those topics must be understood in their historical context. For that reason, we turn now to the origins of privatization.

Three

The Sale Begins — The Origins of Privatization in Hungary

The economic reforms in Hungary, which can be traced to the Kadar regime's New Economic Mechanism and which clearly accelerated in the 1980s, eventually (and perhaps inevitably) led to a well-developed privatization program. Those economic reforms predated the political reforms that led to democratic government in Hungary. Indeed, the privatization program itself began upon the initiative of the Hungarian Socialist Workers party. Ironically, as David Bartlett has observed, the economic reforms of the New Economic Mechanism under the Janos Kadar regime and the reforms of the HSWP under Miklos Nemeth in the late 1980s laid the groundwork for the whatever success Hungary's large-scale privatization program enjoyed.[187] But as one senior government official appointed by the HSWP to oversee privatization said in 1990, "This [privatization] has nothing to do with liberalism. We are simply pragmatic."[188]

The importance of Hungary's privatization program to its development of capital markets can hardly be overestimated. And the groundwork for it clearly was begun under the Kadar regime in the 1960s. To understand fully the role that privatization plays in Hungarian capital reform, one must understand both the history of the process and the history of the organization that has overseen that process. Our story thus begins with the State Property Agency, which since 1990 has overseen the privatization program.

The State Property Agency: Origins, Organization, and Staffing

The offices of the State Property Agency sit in a nondescript office building on Pozsonyi Road, in a crowded section of central Pest surrounded by other office complexes and small shops. The building, only a block and a half from the Danube, is squat and gray, with half a dozen floors. It was built in the 1960s and like other buildings from the same period was clearly built for function rather than aesthetics. It could house any bureaucratic organization anywhere in the world. A security guard sits at the entrance but challenges no one. He seems bored.

Notwithstanding the focus placed on the organization, the SPA occupies only a fraction of the building, and this symbolizes a problem of sorts: the SPA has always been badly understaffed for the massive job it has been given. Indeed, as will be seen, that understaffing largely dictated the course privatization was forced to take beginning in 1991.

The SPA was formed on the first day of March 1990, pursuant to a series of legislation passed by the Hungarian Parliament in January of that year.[189] Passage of the law creating the SPA was precipitated by scandals surrounding the so-called "spontaneous privatizations" that were occurring during the period 1988–1990. In these spontaneous privatizations, which are discussed in more detail later in this chapter,[190] the managers of the state-owned companies initiated and controlled the privatization of their own firms. There arose an accurate perception among the polity and politicians that those managers had a conflict of interest in the process and were in fact using the spontaneous privatization of their firms to profit personally and/or secure their futures in the postcommand economy.[191] Until controlled by the SPA, privatization took place in a "Wild West atmosphere of minimal regulation and outright cronyism and corruption."[192] According to one SPA official speaking in the summer of 1990 about the spontaneous privatizations, "Almost all bids have the same basic problem: the managers started negotiating without an acceptable valuation of the company."[193]

Typically the spontaneous privatizations took the form of joint ventures, with the assets of the state company being contributed to the newly formed joint venture. There were many reasons the joint venture route was preferred by the parties. There was no legal continuity with the old enterprise, so the Western partner did not have to worry about debts or other liabilities of the earlier entity. The joint venture format also facilitated the Western partner's selection of just certain assets for purchase.[194] Of course, these advantages to the Western partner were often detriments to the owner of the old enterprise, the state. After the attractive assets were transferred out, the state was left with a shell holding debts and other liabilities, and unwanted and often unproductive assets; it also found itself in a joint venture. Because, however, the foreign joint venture partner purported to contribute more (i.e, the parties simply valued its contribution at a higher level), the state was generally left the minority partner in the joint venture.[195] The managers of the state-owned company often received equity participation in the newly formed company or lucrative employment contracts. They acted like owners and profited as though they were owners, but they were not owners. One lawyer likened their behavior to that of players in a monopoly game.[196] Questions concerning who owned state assets and who had the power to control their disposition were hotly debated and in need of resolution.[197]

Two well-publicized transactions of 1989 focused attention on the problem. First, in December 1989, the Hungarian hotel giant HungarHotels, which owns two of the most elegant Western-style hotels in Budapest (the Forum and the Atrium Hyatt) as well as many others throughout Hungary, agreed to sell a 52 percent stake in itself to a Western consortium headed up by the Swedish/Dutch group Quintus. The purchase price was set (according to critics, arbitrarily) at $110 million. When it was learned that one of the properties alone could fairly be valued at approximately $60 million, a public outcry ensued.[198] Critics charged that valuable state assets were being sold on the cheap to greedy Western businessmen. The retort argued that the critics did not understand how assets must be valued in a market economy.[199] As will become apparent later in this chapter, these two themes — fear of selling out too cheaply and valuation problems in a postcommand economy much in flux — continued through 1992 to pervade the various debates concerning the proper path of privatization. For present purposes, however, the point is that there ensued a public cry of outrage and demands for reforms in the process. Eventually the transaction was scuttled by the Hungarian Supreme Court on the grounds of gross undervaluation and failure to consider a higher competing offer from Merrill

Lynch,[200] but the thorn had already been planted in the side of the public.

The other highly publicized scandal involved the Apisz chain of stationery stores. In 1989, Apisz negotiated for the sale of itself to an Austrian-led joint venture in a transaction underwritten by Citibank. But the transaction involved so much equity participation by the Apisz management that it could accurately be termed a management leveraged buyout. With the help of their Austrian partners, then, the Apisz management negotiated the price at which it sold the company to itself.[201] Not surprisingly, the purchase price, with its P/E ratio of .33, was generally considered to be unduly favorable to the buyers.[202] In addition, one manager, a former member of the ruling HSWP, received a bonus of approximately $33,000.[203] Again, public outcries could be heard.

Together with the HungarHotels scandal, the Apisz transaction arguably led directly to the establishment of the State Property Agency.[204] According to Zsigmond Jarai, a former head of the Hungarian National Bank and the godfather of Hungarian capital markets, there were two important issues arising out of the spontaneous privatizations that made it important for Hungary to nationalize the process of denationalization of the economy: the question of the locus of authority to sell state assets and the conflict of interest inherent in management negotiated deals.[205]

The process of spontaneous privatization, which resulted in the public scandals, that led to public calls for reform, and eventually led to the establishment of the SPA, can be traced to earlier reform legislation of the Hungarian Socialist Workers party, which controlled the country from the late 1940s until the spring of 1990. In 1984, continuing the attempts at dispersion of economic decision making begun in the New Economic Mechanism, the government enacted the 1984 Law on Enterprise Councils, which transferred some ownership functions of state-owned enterprises from the controlling ministries to enterprise councils controlled by the firms' managements. By 1988 the government (which was struggling under a huge foreign debt and political burblings) seemed resigned to, even perhaps eager for, the economic changes that privatization would bring about. On June 21 of that year, Party Secretary Karoly Grosz told a group of business people in San Francisco, "We would be very pleased if perhaps you would purchase some of our enterprises ... even if they became 100% foreign owned."[206] Six months later, the minister of industry held a press conference at which he revealed a list of fifty-one state-owned enterprises that were available for sale to foreigners.[207] Then the 1988 Act on Associations (the Hungarian Company or Corporation Law), which became effective in 1989, made it possible for state-owned enterprises to form joint stock companies. In July 1989, a new Transformation Law became effective, detailing the process whereby a state-owned enterprise could transform itself into a joint stock company. Together, the three pieces of legislation (1984 Act on Enterprise Councils, 1988 Act of Associations, and 1989 Transformation Act) arguably put management in control of the ownership of the companies and made it possible for shares of ownership to be sold by management. Mechanically, the transaction might involve three steps:

First, the enterprise council would create a new joint stock company, controlled by the managers of the state-owned enterprise. That new company would then issue shares and trade those shares for the assets of the old state-owned enterprise, making the latter a mere holding company. To consolidate its control and facilitate sale to foreign investors, the managers might cause the new joint stock company to sell bonds to the old state-owned enterprise in exchange for the shares. Thus, the state-owned enterprise would be converted to a mere bondholder shell, while management controlled the equity, which could then be sold to foreign investors under terms agreeable to management. In this way, the nomenclatura could trade political capital (currency of the command economy) into economic capital (currency of the new market economy).[208] The possibility for abuse is obvious.

The fact that there were well-publicized incidents of actual abuse led opposition parties to pressure the HSWP, which still clung to power, to control the process of spontaneous privatization.[209] By early 1990, after the HungarHotels and Apisz incidents, the ruling HSWP had already set open democratic elections for that spring. Control of the privatization process became an important point of debate. On one side were the Alliance of Free Democrats (Szabadsag Demokratak Szovetseg, or SDS) and the Alliance of Young Democrats (Fiatal Demokratak Szovetseg, or Fidesz), who argued that putting the economy quickly into private hands was the most important goal. On the other side were the conservative parties, most prominently the Hungarian Democratic Forum (Magyar Demokrata Forum, or MDF), which argued that the process had to be tightly controlled to avoid selling off valuable Hungarian assets too cheaply to rapacious foreigners.[210]

Meanwhile, the spontaneous privatizations were increasing in number in a political and legal vacuum. The HSWP still held power, but from June 10, 1989, when it announced its willingness to negotiate with nine opposition parties for free elections, it was clearly a lame-duck party. The legality of the spontaneous privatizations was dubious, but there were no legal constraints; thus the Hungarian maxim "That which is not forbidden is permitted" applied. There existed an Office of State Commissioner on Privatization that issued guidelines and directives, but it acted only in an ad hoc fashion and was often overruled by some other governmental organ claiming jurisdiction.[211] According to an SPA spokesman, speaking of spontaneous privatizations that took place before the SPA was established, "In some cases, management ravaged the companies. Management took advantage of the [political and legal] vacuum."[212]

The MDF would win a plurality of the vote in the elections and form a coalition with other conservative parties, but the HSWP government did not wait before acting to control the process. January 1990 saw a spate of economic reform bills pushed through by the Nemeth government.[213] Included were the two cornerstones of the controlled privatization program that would oversee the process. They were Law VII of 1990 on the State Property Agency and the Management of State Property Operating in Enterprises, and Law VIII of 1990 on Protection of State Property Entrusted to Enterprises.[214] Together those two laws established the State Property Agency as the title holder of state-owned assets and established controlled mechanisms whereby state enterprises and state assets could be privatized under the auspices of the SPA. Small transfers of state-owned property, i.e., worth less than the equivalent of approximately $300,000, were exempted from SPA oversight for practical reasons.[215] So were sales of minority equity stakes.[216] Nonetheless, as of July 1992, the SPA controlled 98 percent of state assets.[217]

The SPA formally came into existence in March 1990, and from that time it served two functions: management of the state's portfolio of assets and privatization of those assets.[218] Seed money for its establishment came largely from Western sources. The World Bank contributed $66 million, and the United Kingdom paid the salaries of two British advisers out of the 25-million-pound "Know How Fund for Hungary," established in 1989.[219] Western capitalists generally welcomed the establishment of the SPA, since under the Nemeth government the process was confusing and chaotic.[220] Their enthusiasm was much moderated by their experience with the SPA over its first two and a half years of life, however.

The SPA was overseen by a board of eleven directors, seven of whom came from inside the government and four of whom came from outside it. They were appointed by the prime minister for five-year terms but could be dismissed at any time. Day-to-day operations were run by a managing director.[221] The organization as a whole came under the jurisdiction of the Council of Ministers State Audit Office.[222]

As of late 1992, there were 12 departments within the SPA, each with its own area of specialty. Each was headed by a department head, and within the department the department heads maintained a great deal of latitude to exercise discretion in decisions concerning acceptance or rejection of terms of individual transactions. But each week, for approximately three hours, transactions would come up for approval by the board of directors. No transaction could be finalized without approval of the full board.[223] Thus, notwithstanding the complaints about bureaucratic delays within the SPA (discussed below[224]), organizationally there has been little to distinguish Hungary's State Property Agency from large business organizations. Delays and red tape were always in the execution, not the organization, of the SPA's work.

One of the most substantial barriers to effective and rapid privatization of eligible state-owned enterprises has always been the woeful understaffing of the SPA. It was formed in March 1990, but by October of that year it still had a staff of only 40.[225] At that time, the SPA was claiming it expected to be able to review bids for firms being privatized within three weeks.[226] Given its staffing level, it was not surprising that evaluations often dragged on for months and almost never were completed within the three week period.

By the end of 1990, the staff at the SPA numbered 50,[227] and they were still, using the term of an American accountant, "swamped."[228] So the SPA added staff rapidly. In October 1991, the size of the SPA staff had risen to 130, more than threefold over its size the previous year. Still, as 1992 dawned, the managing director of the SPA was telling reporters that the SPA lacked the funds to employ a sufficient number of people, which led to casework overload and concomitant delays in the process.[229]

By November 1, 1992, the SPA had approximately 250 employees and was still growing. According to sources within the SPA, the problem was not so much with the number of bodies working at the SPA; rather it was a problem of quality of staff. The financial reviews required by the SPA would entail substantial expertise even in a purely market economy in which the principal actors all spoke the same language. But in Hungary, analyses were complicated greatly for two reasons: (1) economic performance was distorted in the command economy, making it almost impossible to gauge the relevance of performance figures stated by the Hungarian firm being privatized, and (2) market valuations for assets were terribly difficult to make when the purpose of the sale of state assets was to develop precisely the market that would make the valuation possible. Finding sufficient numbers of Hungarians with the financial sophistication necessary to understand basic finance concepts and the unique position of the country's former command economy would be difficult even without the complication of language difficulties. Insufficient numbers of people at the SPA with the rudimentary finance skills have sufficient language skills to deal with bidders and economic advisers who speak German, English, Japanese, Italian, and French.[230] This combination of lack of business acumen and language skills resulted in backlogs and undue caution born of uncertainty. Thus the understaffing at the SPA is best understood as a problem that would have to be solved on the quality front rather than the quantity.[231]

To understand the origins of Hungary's privatization program, we must now explore the motivations for that privatization. As we see, these motivations have resulted in the unique style that privatization has taken in Hungary.

Why Privatize? And How Should It Be Done?

Trying to figure out exactly how much property the Hungarian state had available to privatize has always been a tricky business. The problem is caused partly by the fact that the question "Who owns this?" was not a particularly relevant question when ownership did not equal control. The state was in control of over 90 percent of the means of production in the late 1980s, and the Hungarian

Socialist Workers party was in control of the state. Because control was centralized, it was not necessary to delineate precise ownership interests. As one official put it, "The problem is, we don't know exactly who owns these companies. The State is the owner, but who is the State?"[232] Local councils, ministries, cooperatives, enterprise councils, and individuals often had indistinct ownershiplike interests in specific assets. For the Hungarians, beginning to privatize was a bit like starting to set up a garage sale and realizing that one was not sure to whom all those items in the garage belonged. The Hungarians solved that problem by calling virtually all state-controlled property "state property" and leaving for another day the question of how to divide the proceeds of its sale.

A further difficulty came in estimating the value of that property. One could capitalize the earnings generated by the assets, but the earnings figures were distorted by the command economy. As one Austrian banker put it, "The old command system produced meaningless figures."[233] In addition, past figures were of little predictive value in the turbulent years 1988-1992. Future earnings were almost impossible to predict given the instability of the Hungarian economy and the fact that earnings potential depended on whether, when, and to whom the assets would be privatized.[234] One could try to assign market values to those assets, but since market value is by definition what a willing buyer would pay to a willing seller, that was a terribly elusive figure: no one knew ex ante whether there would be any willing buyers at all. The idea of market values had very little meaning, since the whole idea of the process was to *create* a market.[235]

These problems notwithstanding, there were floated about various estimates of how much the Hungarian government had to privatize, and the figures were impressive. The scope of the task before the Hungarians was unique and daunting.

By most accounts, there were in mid-1990, approximately 2000-2400 state-owned enterprises in Hungary,[236] accounting for approximately 85-95 percent of economic output, depending on how that output was expressed, although some estimates varied from those ranges.[237] These figures do not include the small retail shops and restaurants (numbering perhaps 10,000) that were relatively less important macroeconomically and which were to fall under a separate privatization program, as discussed later in this chapter. The value of those enterprises when the process started was estimated at between

$26.7 billion[238] and $30 billion.[239] Because, however, these estimates are based on book value rather than market value and did not include the value of intangibles, the market value of the state-owned property may have been grossly overestimated or underestimated. For example, the book value of plants and equipment of moribund state firms whose primary market was lost when the Soviet Union and COMECON collapsed[240] probably overstated market value. In addition, under Hungarian accounting rules, Hungarian firms under depreciated assets for years.[241] And yet land values were carried on the books of Hungarian firms at just fractions of what some Hungarians thought Westerners would probably be willing to pay.[242] Only later did officials realize how badly they had overestimated the market value of state assets.[243]

Although it was unclear how much Hungary had to sell, most observers felt confident that most of the state assets should be sold. The advantages to be gained from putting the majority of productive assets into private hands were thought to be numerous. They fall into three broad groups. The first was political: it was widely acknowledged that a strong private sector was likely to create a sufficiently diffuse middle class necessary to sustain the democratic reforms put in place.[244]

The second goal was the reinvigorating of the economy. Gyorgy Matolcsy, a privatization specialist and government economist in Hungary put it this way:

> Privatized firms will invest more readily in new technology and equipment, they will bring in better know-how and management skills and will organize themselves more efficiently than their state-controlled predecessors. They'll encourage entrepreneurship and incentive, generate profit and get the economy moving.[245]

Or, as the *Economist* succinctly noted in April 1990, "Ownership change is the key: real improvement will not come about until genuine private owners take responsibility for the inefficient and nomenclatura-dominated economy."[246] And indeed in the first half of 1990, the privatized enterprises showed substantial growth, whereas the state sector as a whole lagged.[247] Of course, one must not make too much of such a datum; previously successful firms were the obvious first candidates for privatization.

Finally, the Hungarian government saw the sale of state assets as an important source of funds. One use would be the repayment of Hungary's onerous $20 billion foreign debt, which in the late 1980s and early 1990s was the largest per capita in Europe.[248] Another was to offset low government revenues that had created a worrying and worsening budget deficit problem by the early 1990s.[249]

While these three goals were crystalizing into government policy, a large number of proposals were being discussed inside and outside Hungary as to the best methods to employ in privatizing state-owned enterprises. Some commentators were pushing giveaway schemes. Professor Jeffrey Sachs of Harvard University had successfully persuaded the Poles that public distributions of state property were a better idea than sales of that property primarily to foreigners, and he suggested that as a model for other former Socialist countries.[250]

A variation on that idea involved making installment sales of transport equipment and small agricultural holdings to the users of those means of production, with the payments collateralized by the equipment and/or real property. This procedure would then be followed by putting all other state-owned assets into large holding companies and distributing the shares free of charge to the public.[251] Similarly, economists Blanchard and Layard suggested putting all state-owned enterprises into five holding companies, and then distributing the shares to the public without cost. Up to 20 percent of the shares could be resold per year, and the firms within the holding companies would be privatized over a ten-year period.[252] And the Independent Smallholders party, which joined the ruling coalition after the 1990 elections, advocated widespread redistribution of state property, particularly agricultural land, to the people.

The giving away of state property had political and economic allure. The state's acquisition of the property in the 1940s and 1950s was widely viewed as illegitimate in the first place. And such a method would result in rapid privatization and the immediate creation of an ownership class. But there were problems.[253] One involved the perplexing but not insurmountable technical difficulties. Who would get how much? How would one administer such a program? Similar steps undertaken in Poland were described in early 1990 by government economist Laszlo Antall as "radical" and "chaotic."[254] In addition, giveaways would not allow the state to use revenues for debt reduction. Since much of the debt was accumulated to finance a consumer economy much better than that enjoyed by Hungary's

Socialist neighbors, perhaps it would be fair for the public to forego distribution of the property in favor of sales that would repay that debt. A third problem was the fear that diffuse ownership would permit greater control by insiders, who would accumulate small but important blocks. Fourth, macroeconomic stabilization efforts (particularly attempts to control Hungary's 30+ percent inflation rate) could be hindered by a surge in public wealth. For perhaps these and other reasons, Hungary rejected what one commentator labeled a "19th century small owner solution."[255] And in 1990 the Hungarian Constitutional Court ruled against a claim that property *must* be restored to the former owners.[256]

Other ideas, similarly based on notions of historical justice and a desire to effect rapid widespread property distribution, involved distributions to employees, perhaps through the use of employee share ownership plans (ESOPs).[257] Indeed, during the campaign leading to the election of 1990, the Christian Democrats and the Social Democrats both advocated privatization primarily through the use of ESOPs.[258] Both the SDS and Fidesz were suspicious of such moves as smacking of "third way" circumventions of true market reforms.[259] Although ESOPs were in fact used eventually on a small scale, the idea of widespread gratis distributions to workers was rejected by the government for many of the same reasons that resulted in the rejection of public giveaways generally.[260]

One idea that, unlike free public distributions, would have aided the Hungarian foreign debt situation was the debt-for-equity swap: give state-owned property to foreign creditors in exchange for the debt.[261] With a foreign debt of $20 billion and a total value of state-owned enterprises estimated at $30 billion, this was never a viable comprehensive solution for political reasons: foreign creditors would end up owning two-thirds of Hungarian assets. As government economist Laszlo Antall put it: "We want to avoid swaps. We saw how they affected Mexico and Bolivia."[262]

Other ideas focused on the possibilities of creating a system of institutional ownership. For example, nonprofit domestic institutions could receive grants of shares, thereby simultaneously unloading the state-owned enterprises and alleviating the budgetary problems of the government by cutting those recipient institutions free from the need for state support.[263] Another idea involved the exchange of equity among the various state-owned enterprises, creating a system of mutual intercorporate ownership.[264] These plans were not sufficiently different structurally from what already existed ever to warrant much attention.

After the privatization process was in place, other creative ideas sprang up. For example, in early autumn of 1991, the Hungarian National Bank announced plans to float a fixed dollar Eurobond offering that would include warrants to purchase privatized firms. The idea promised to save the bank approximately 3 to 4 percent on the bonds' interest rate and would guarantee the purchasers' ability to buy shares in the privatized firms at an attractive price. The plan was eventually scuttled, in part because the bank was proposing to issue warrants for the purchase of shares in companies that were not yet privatized.

Nonetheless, the early 1990s was a time of terrific creativity on the part of the Hungarian privatizers. They did not hesitate to continue proposing new and unusual adjustments in the planned privatization methodologies as conditions warranted. For example, as sources of foreign capital became more scarce and concern over low levels of domestic participation in the privatization programs became more pronounced, ideas such as low interest loans for domestic investors, privatization credit cards, domestic share set-asides, asset leasing plans, and asset management schemes were floated, some of which became government policy. These midcourse adjustments are discussed more fully below.[265] The point here is that in the late 1980s and early 1990s, privatization could have taken many different paths. The future was wide open, and Hungarian policymakers were unfettered by history and precedent but were also inexperienced and were exploring untraveled terrain.

Once the government took control of the privatization process in 1990 and that process began to reflect government goals for it, the three motivations discussed above[266] (entrench political reforms, revitalize the economy, and raise revenue for the state) deeply informed the methodology the government would employ. For that reason many of the ideas for how best to privatize were rejected, particularly those that would not result in any revenue to the state. Had the government not put so much emphasis on raising revenue, it might have distributed state assets to the population quickly (as Poland and Czechoslovakia did[267]), thus rapidly creating a private economy with wide domestic wealth dispersal.[268]

But the first privatizations were uncontrolled and therefore not rationally related to any conscious government policy.

Setting Out: Pre-SPA Attempts at Privatization

The sales of state-owned property to private business and individuals before the January 1990 legislation discussed above[269] can, for convenience, be broken down into three categories: pre-SPA private placements, pre-SPA public offerings, and spontaneous privatizations. Notwithstanding some successes, in retrospect it seems clear that the publicly perceived difficulties of these pre-SPA privatizations led to the government's nationalization of the privatization process and the establishment of the SPA.

Private Placements: The Examples of Tungsram and Graboplast

By the late 1980s, the huge Hungarian light bulb manufacturer Tungsram produced more light bulbs than any other manufacturer in Europe. It possessed a commanding market share within the COMECON countries and had a well-developed market in Western Europe as well. Nonetheless, it was badly managed, and its financial difficulties were becoming severe. Those difficulties seemed destined to increase, since the manufacturing facilities were not being updated as they should have been. In 1989 the Hungarian Credit Bank was holding most of the debt of Tungsram, and it began pressing the company. After a series of negotiations, the parties (both of course controlled by the state) agreed that the bank would take a majority equity stake in Tungsram in exchange for the debt.

But the Hungarian Credit Bank was not at all interested in holding as an investment a majority equity stake in a sinking industrial ship. So in April 1989 it sold 49.5 percent of Tungsram's equity to an Austrian consortium headed by Girozentrale, an Austrian investment banking firm, for $110 million. In November, General Electric bought Girozentrale's stake and enough other stock from the Hungarian Credit Bank to give General Electric 50 percent plus one share. The total price to GE was $150 million, and Girozentrale netted a handsome $40 million profit without putting anything into the company. Many cried foul. The general public view was that the slick Austrian financiers had bought sound Hungarian assets for far too low a price. Many members of the public and press were left wondering what the persons negotiating on behalf of the seller had gotten in the deal.[270]

But the critics were missing some things. Girozentrale had purchased the equity of a huge industrial firm, employing thousands, that was on the verge of economic collapse. It took the risk that the firm could be sold. The gamble happened to pay off, but Girozentrale had to carry Tungsram for seven months while it scrambled to find a buyer. If no buyer had been found soon, Girozentrale would have lost tens of millions of dollars. And in the end Tungsram was saved by the transaction. According to George Varga, appointed by General Electric to run Tungsram, the company would not have survived 1990 without GE's immediate investment of $20 million.[271] GE put another $20 million into the company in 1991, securing Tungsram's future as a contributor to the Hungarian economy. Viewed as a whole, the transactions was tremendously favorable for Hungary and very risky for Girozentrale. Nonetheless, policy changes are impelled by perceptions not by facts, and what was lauded in the Western press as a successful and massive door-opening by a U.S. industrial giant[272] was in some Hungarian circles a scandal.

But not all pre-SPA private placements were controversial. Perhaps the most successful was that of Graboplast. Graboplast was a manufacturer of floor and wall coverings and imitation leather. Its reputation was that of a quality producer, and it was generally regarded as being well managed. Neither of these attributes was a pervasive phenomenon among Socialist industrial companies. In 1989 the company had sales of approximately $71 million, 30 percent of which were exports and 25 percent of which were exports to the West. Its 2000 employees were split between two factories. Its problem was that it needed an infusion of capital. It wanted to update its marketing processes, improve its ability to purchase supplies on short notice, and update its data processing and warehouse and stock control. Unfortunately, it was already heavily burdened by debt and could therefore neither borrow more funds nor use its cash flow to finance modernization.[273] Graboplast's management turned to the investment banker/brokerage firm CA-BB, a joint venture between the Austrian firm Creditanstalt and the Hungarian commercial bank Budapest Bank. CA-BB arranged a private placement of about a third of Graboplast's equity to Western investors, which provided an infusion of approximately $12 million. Graboplast used that money to pay down its expensive debt, thus freeing up cash flow for its modernization plan.[274] The transaction did not close until after the SPA was established, but because the transaction was for a minority stake, no SPA approval was required.

Pre-SPA Privatization: The IBUSZ Public Offering

Not all the privatization originating before the SPA began its work consisted of private placements, and unfortunately for the principal movers of the IBUSZ public offering, not all privatization transactions were as warmly received by the Hungarian public as was the Graboplast transaction.

In 1989 the state-owned company IBUSZ was the largest travel agency in Hungary, with approximately 50 percent market share in the country.[275] It and its investment adviser Girozentrale planned to make a public offering of approximately one-third of the IBUSZ equity in Hungary and also on the Vienna Stock Exchange. Eventually this figure would be raised to approximately 40 percent[276] and then to 42.7 percent.[277] When it became clear that the Budapest Stock Exchange would reopen sometime in 1990, IBUSZ coordinated the timing of the offering to coincide with the BSE's opening. The simultaneous events were expected to make a big splash. The sale of the stake was expected to bring in approximately $18 million, part of which would go to the state and part of which would be used by IBUSZ to expand its operations. Said IBUSZ president Ericka Szemenkar: "Privatization is really going to help our company. We need the money it is going to raise for us. We have big plans for the future."[278]

This partial privatization of IBUSZ was almost fully planned by the time the SPA began its work in March 1990, and the offering was made on June 21, 1990, the day after the BSE opened.

The euphoria was extraordinary. The offering was oversubscribed 23-fold.[279] The stock opened at the offering price of 4,900 Hungarian forints per share (approximately $65) and immediately leapt that day to 8,000 HuF. Two days later on June 23, the stock was quoted at 12,500 HuF before it settled back to 10,000 HuF on July 6.[280]

A political row ensued, with the public and members of Parliament complaining that given the subsequent run-up in price, there had been a "selling out of Hungarian national property" on the cheap.[281] Although the transaction was planned primarily by IBUSZ management and its investment advisers at Girozentrale, the political heat fell primarily on the managing director of the SPA, Istvan Tompe, who had come to head the agency just before the ruling coalition formed by the MDF took power following the elections in the spring of 1990. MDF members of Parliament called on Tompe to resign. Leading the attack was MDF member of Parliament Istvan Bethlem, who labeled the transaction "scandalous" and the SPA "incompetent."[282] Ironically, only weeks before, Bethlem had favored a private placement through Creditanstalt in the 2,500–3,500 HuF per share range, substantially less than the initial public offering price for the IBUSZ shares.[283] Tompe lamented to a reporter, "Two cases like this and all the interest in investment in Hungary will disappear."[284]

Much of the flap surrounding IBUSZ was politically motivated. The real question was not whether the price was set too low, but rather, who should head the SPA. Tompe had been elected to his post by an all-party committee in March 1990, just before the MDF coalition assumed control of the government. He was widely respected in the financial community of Budapest, but Prime Minister Jozsef Antall's chief economic adviser Gyorgy Matolcsy wanted him out.[285] The uproar bore the markings of a well-orchestrated political smear by the MDF.

If those who set the initial offering price for IBUSZ were guilty of anything, it was underestimating the irrational demand for IBUSZ shares among foreigners, not in understating the intrinsic value of the firm and its future prospects. Based on financial statements audited by Price Waterhouse, Girozentrale chose an offering price that put the P/E ratio, based on projected 1990 earnings, at 9.2,[286] not unreasonably low for an offering in a country with a brand new democratic government for the first time in nearly fifty years and a shaky economy. And subsequent performance of IBUSZ shares seemed to vindicate the conservative valuations made by Girozentrale. A little more than a year later IBUSZ shares could be purchased for only 2,800 HuF per share, a discount of almost 40 from the initial offering price.[287]

Nonetheless, a month after the public offering, Tompe was ousted and the MDF prime minister Jozsef Antall replaced him with Lajos Csepi, who continued to head the agency into the early 1990s.[288]

His appointment was seen by financial analysts as a sign that the government wanted to slow down the pace of privatization. If true, the government got its wish. Tompe had been quoted as saying he wanted to privatize 80 percent of the economy within a few years.[289] Csepi turned out to be much less ambitious. The IBUSZ/Tompe affair caused some to wonder later whether the snail's pace of privatization that ensued was really a result of intractable problems (as the government suggested) or a conscious government policy designed to pacify reactionaries in its ranks who were not eager to privatize the economy at all. One fear of the critics of privatization was realized in the offering: of the 42.7 percent offered to the public, only 4.2 percent was purchased by Hungarians; the rest went to foreigners.

Eventually it turned out that the SPA's biggest mistake was selling only 42.7 percent of IBUSZ at 4,900 HuF per share. It retained the other 56 percent as an investment (the other 1.3 percent going to employees) with the idea that it could later be sold in a secondary offering at a higher price once the economic benefits of privatization accrued to IBUSZ.[290] The SPA decided in late 1991, however, to sell its 56 percent block in a competitive tender; the price languished below 3,000 HuF per share.[291] The SPA had hoped to dump its shares on the Budapest Stock Exchange, but exchange officials objected, complaining that the effect would be a further weakening of IBUSZ share prices and the exchange generally.[292] The SPA relented and agreed to the competitive tender. There was a delay, however, because the IBUSZ charter prevented any one shareholder from exercising more than 5 percent of the voting rights. This provision had been inserted to keep a foreign shareholder from snapping up control of IBUSZ in the course of public floatations of the stock.[293] On February 26, 1992, the IBUSZ shareholders were forced to remove that restriction from the charter to permit the SPA to sell its shares in blocks.[294] It did so that spring, selling a 49 percent stake to the Hungarian Commercial Bank (itself owned 44.2 percent, directly or indirectly, by the state[295]) and 7 percent to the State Social Security System.[296] The SPA then bought back 66 million HuF worth of the shares to make them available for compensation coupons to those whose property had been unlawfully appropriated by the state in the 1950s.[297]

Not surprisingly, the IBUSZ offering had a souring effect for many participants. The public perceived it to be a scandalous sellout of their property to foreigners. It cost Istvan Tompe his job. Investors in the shares lost vast sums of money. And perhaps worst of all, a majority of the shares were still held either by state organs or by organizations at least indirectly controlled by the state.

Although the IBUSZ deal was incubated in the pre-SPA era, by the time it became a scandal, the SPA was up and running, so any policy reactions were limited to adjustments of the state-controlled privatization program that had been put in place with the January 1990 legislation. It was other scandals, including the Tungsram transactions chronicled above,[298] that actually led to that state control. We turn now to the system of spontaneous privatization generally, to which the state-controlled system was a reaction.

Spontaneous Privatization

This chapter has already touched on the subject of spontaneous privatization, noting the three pieces of legislation (the 1984 Law on Enterprise Councils, the 1988 Act on Associations, and the 1989 Transformation Law) that made spontaneous privatizations possible. The chapter has also recounted the examples of HungarHotels and Apisz and the cries for reform that followed those scandals. Those examples, however, offer an incomplete picture of the management-initiated privatizations between 1988 and early 1990.

Between 1987 and 1990, approximately 150 state owned companies, representing approximately 8 percent of all such companies, spontaneously privatized themselves.[299] The economic effect may have been even greater; at least one economist has suggested that more than 8 percent of GDP was privatized in that period because those that privatized themselves were relatively more economically significant than those that did not.[300] This figure is difficult to verify and may overstate the percentage of state assets spontaneously privatized, since by August of 1991, only approximately 5 percent of state assets (measured by book value) had been privatized, and some of those privatizations were SPA-directed.[301] Still, it is a fact that there were many spontaneous privatizations before the SPA came into existence, and their size and numbers made them economically significant. Indeed the numbers just cited may actually underestimate the degree to which assets were being privatized spontaneously in the period. By one estimate, during 1988 and 1989, 1,600 new companies were formed, pursuant to the new Act on Associations, using (at least partially) state assets.[302] We might call these "spontaneous partial privatizations."

The first reported spontaneous privatization was that of Medicor in 1987. Medicor was at that time the largest Hungarian producer of medical instruments. It reportedly reorganized its ten factories into joint stock companies and sold them off to Hungarian investors, banks, and companies.[303] How it managed to do so before the 1988 Act on Associations and 1989 Transformation Act made it possible to transform state-owned enterprises into joint stock companies is a mystery.

Others would follow. For example, it was during the pre-SPA period of spontaneous privatizations that Schlumberger Industries acquired a 75 percent stake in Ganz Electric Motors and Hunglet acquired 51 percent of Ganz Engineering. Most such acquisitions took the same form. The privatizing firm would contribute a majority of its assets to a joint venture. The partner would then contribute assets valued by the parties at more than the value of those assets contributed by the privatizing firm, thus assuring the acquirer of control of the joint venture and leaving the state holding assets the acquirer did not want and a minority position in a joint venture.[304]

Not all the spontaneous privatizations during the period, however, involved hints of scandal or impropriety. One notable success story is that of Janos Petrenko. In 1989 the huge steel complex at Ozd was in serious financial difficulty as a result of poor management that had led to a failure to modernize and crushing debt loads. When the complex was about to collapse, Petrenko was able to buy a piece of the facility in spontaneous privatization. He modernized his small steel mill, made it profitable, and saved the jobs of at least a portion of the Ozd workforce.[305] Unfortunately for Hungary, few individuals had the financial means of a Janos Petrenko, and Hungary would have to rely largely on foreign capital in its privatizations.[306]

Structurally, there was never anything wrong with the spontaneous privatizations. Even where the state is left with a shell company that holds a minority position in a joint venture, there is no problem so long as the transaction is negotiated fairly and at arm's length, thus ensuring to the greatest extent possible that the state-owned enterprise (and therefore the state) would get an appropriate percentage interest in the joint venture. Unfortunately, there were several problems that prevented the state from realizing what it and the public considered fair value for the assets it was either selling outright or contributing to joint ventures.

The Origins of Privatization in Hungary 49

Two of those problems were discussed briefly earlier in this chapter. The first was that it was never clear that the managers of the state-owned enterprises had any authority to sell state assets,[307] and it was often even unclear who owned the assets being sold.[308] The second was that there were opportunities when management was negotiating with the purchaser for management to profit personally, whether by taking lucrative employment contracts, taking equity, or whatever.[309] That gave them a conflict of interest in the transaction and led to suspicions, as in the Apisz and Tungsram privatizations discussed above,[310] that management had agreed to sell below value in exchange for a personal benefit.[311] As the managing director of the SPA would later tell a reporter, "We strongly suspect that in many cases the transformation has served the interest of those managers who wanted to save themselves and their influence in the future."[312]

Three other problems with pre-SPA privatizations are worth mentioning here. First, in the pre-SPA privatizations, there was no provision for any proceeds of sales to go into the state budget.[313] The deals were almost all asset sales, usually asset contributions to joint ventures dominated by the acquirer. If cash flowed to the selling enterprise, it generally went to a subsidiary of the enterprise, meaning that the enterprise itself (which continued to be owned by the state) was merely a holding company. If the enterprise itself received the money, it was kept by the enterprise and not turned over to the state.[314]

This phenonmenon whereby the state's interest was converted from an economic enterprise to a holding company or a much smaller firm with a substantial infusion of cash theoretically should not have affected the value of the state's interest at all. But the phenomenon turned out to be a problem in practice. The cash often lined the pockets of the managers.[315] If a subsidiary was holding the acquirer's payment, the subsidiary was often liquidated by the management.[316] There was no formal liquidation, of course, but assets would dissipate over time as managers prepared for new lives in a market economy.

Finally, there was a widespread perception that, regardless of malfeasance on the part of firm management, the state owned enterprises were selling themselves or their assets far below fair value. As one Hungarian economist pointed out in 1990, the practice of spontaneously privatizing by contributing state enterprise assets to a joint venture was fine for the state if the state enterprise contribution was valued fairly; often, according to this economist, it was not.[317] The problem was not necessarily graft on the part of managers; it may have been ignorance. Managers had never had to put market values on companies before, and there were no guidelines to ensure proper valuation.[318] An observer has suggested that "the truth is most Hungarians wouldn't know a good price from a bad one."[319] But in fairness to the Hungarian managers, even Westerners were having trouble putting values on the concerns. They could assess past profitability and cash flow, but they had a difficult time measuring the effect of central planning and fixed prices.[320] When the Soviet Union folded up its tent and COMECON dissolved, it made the future even more uncertain than it otherwise would have been. From 1989 to 1990, the Soviet Union's percentage of Hungarian foreign trade went from 30 percent to 14 percent[321]; COMECON disbanded in 1991.

Or the problem may not have been a problem at all but a case of unrealistic expectations on the part of the Hungarians. One Western financier put it this way: The Hungarians want unrealistic prices for their companies. They have to understand that there are other places to put money. And if I don't get a 20 percent to 30 percent discount in Hungary to make up for the difficulties of doing business here, it's not worth it.[322] Another observer echoed those sentiments, noting, "The desire to get top dollar has caused [the Hungarians] to turn down things where the valuation is not realistic."[323]

As in the case of the HungarHotels privatization discussed above,[324] many Hungarian observers have had a difficult time understanding the difference between "fair value" from an accounting standpoint and "market value" in an economy fraught with uncertainty and in-country risk.[325] One Austrian banker said in early 1991 that he looked at the value of similar assets in the West and divided by five to arrive at their value in Hungary.[326]

Although the abuses of spontaneous privatization were not government policy, neither were they occurring behind the scenes. Indeed, in the political vacuum of the late 1980s in Hungary, when it was clear that the HSWP headed a lame-duck government, there was very little government policy at all. The caretaker government (and therefore privatization policy) was adrift in a chaotic Wild West atmosphere.[327] Privatization was an uncontrolled free-for-all under the Nemeth government.[328] In effect, the government told the managers of state-owned enterprises that it was in favor of privatization and wanted the managers to privatize the way they thought best.[329] The government then simply adopted a policy of noninterference.[330]

These problems attending the pre-SPA spontaneous privatizations were not inherent in their being management driven. Indeed, the conflict of interest the managers found themselves in when negotiating for the sale of their firms' assets is precisely the conflict of interest Western company managers find themselves in when negotiating similar transactions. The problems came about because no one was monitoring management to make sure it was not engaging in those practices in which it had an incentive to engage. The privatizations were not motivated exclusively or perhaps even primarily by private wealth-seeking by the managers but rather by the desperate economic condition of most of the firms involved. Banks were calling in the loans.[331] But once a decision to privatize was made, one can understand the incentive for managers to seize the opportunity. Most of the managers were former (or soon to be former) Communists and very much part of the nomenclatura. Their ideology was discredited, and the financial troubles they found themselves in as managers indicated that they would be viewed later as poor managers as well. In the new democratic, market-economy world, how were they going to find positions of equivalent prominence and affluence to what they had been enjoying? They were told to fend for themselves without governmental control. We should not be surprised, let alone shocked, to find that many of them attempted to secure their futures "on the other side" through the sale of the enterprises they had been managing.

As it turned out, spontaneous privatizations continued with success after the SPA started controlling the process.[332] The abuses were eliminated for the most part, and spontaneous privatization (with its emphasis on decentralization and speed rather than statism and rationality[333]) was the most successful and popular route to privatization in the early 1990s.[334] In 1990 alone, there were 230 instances of spontaneous privatization, and once the SPA was involved, they tended to be scandal-free.[335] By August 1991, approximately 5 percent of all state-owned property had been privatized that way;[336] by late that year the figure was approximately 6 percent.[337] As one government economist put it, "this so called spontaneous process does work."[338] Managers could continue to take the lead in seeking privatization partners and could conduct the negotiations on behalf of their firms. Indeed, the SPA repeatedly voiced its encouragement of this privatization route.[339] But the transaction had to be approved by the SPA, which was concerned only with ensuring the possibility of competitive bids and what it considered realistic valuations for the company.[340] As we will see below, the SPA probably put too much emphasis on its valuations and competitive bidding, thereby slowing up the process and chasing away some purchasers unwisely.

The cleanup of the spontaneous privatization process had to wait for the SPA's establishment in 1990. One thing was clear by then: the economic and political climates were changing rapidly, and the government had to choose a direction for privatization and rationalize a process that was becoming sufficiently scandal-ridden by the spontaneous privatizations to threaten the political support for the reforms. The route chosen — sales of companies and assets controlled by a government agency — was dictated largely by the motivational mixture for privatization that survived the election of 1990. The government sought to create a private economy while at the same time bringing in money to the state coffers.

Four

Privatization Under the State Property Agency — The Opening Programs

Although the State Property Agency was created before the democratically elected MDF coalition came to power in 1990, the MDF government moved quickly after the election to assert control over SPA policies. The first head of the SPA, appointed under the old HSWP government, lasted only five months and was gone within weeks of the MDF coalition's ascendancy. The SPA privatizations under the MDF government evidenced, on the one hand, a combination of pragmatism and flexibility. On the other hand, those privatizations evidenced a steadfast commitment on the part of the government to certain basic principles. When we examine the privatization programs under the SPA, therefore, we see a tension between pragmatism and principle.

The basic principles adhered to by the MDF government in its privatization policies can be grouped as follows:

1. State assets would be sold rather than given away.
2. Care would be taken to ensure a "fair" price, even if that meant the process would move slowly.
3. Foreign participation would be strongly encouraged (indeed relied upon), but attempts would be made to facilitate the participation of domestic Hungarians when possible.

These basic principles put the SPA in a crucible of political debate. Poland and Czechoslovakia were giving away assets to the public, and there was pressure from the far right (particularly from coalition partner, the Independent Smallholders party) and from certain academic circles (primarily Western) to follow that route. The main opposition parties, SDS and Fidesz, together with both Western and Hungarian financiers, argued that the SPA was being far too cautious in its attempts to secure a "fair" price for assets and would be better off moving quickly to create a private economy. Finally, the dominance of foreigners in the process caused consternation among the public. We see these principles expressed repeatedly in the policies of the SPA despite frequent pragmatic course changes.

In 1990 and 1991, the SPA embarked on five specific privatization programs, working at both ends of the spectrum with respect to size of enterprise. We take them up one at a time in this chapter.

The First Privatization Program: Selling the Crown Jewels

The so-called First Privatization Program (FPP) (sometimes called Phase I of the SPA's Active Privatization Program[341]), involved a hand-picked selection of the large enterprises that were thought to be most attractive to Western investors. The FPP was a huge flop. The SPA had already gotten off to a bad start when its first privatization candidate, IBUSZ, became mired in controversy immediately after the public offering in June 1990. But the SPA's problems were more pervasive than that, a fact that became clear over the next year and a half.

The SPA announced the FPP with much fanfare in September 1990. For the few months preceding that announcement, there was a lot of speculation in the press and in Budapest financial circles concerning which companies were going to be privatized first.[342] Western investor interest was nearly euphoric. And a major obstacle to Western investment was removed when the Constitutional Court ruled against the notion that former owners' claims took precedence over privatization.[343]

In its announcement, the SPA stated nine aims of this first program:

1. Make the process transparent to all, thus generating public support
2. Ensure an active government role

TABLE 4-1: CANDIDATE COMPANIES OF THE FIRST PRIVATIZATION PROGRAM
(figures in $ million)[347]

Company Name	Book Value	Sales	Pretax Profit
Centrum Dept. Stores	69.6	248.3	5.33
Danubius Hotel and Spa Co.	88.5	52.0	7.6
Forest Machinery Producing Co.	2.0	2.2	.02
Gamma Works (medical and other high-tech equip.)	23.4	21.3	.67
Hollohazi Porcelain Works	6.5	5.7	.51
HungarHotels	145.8	100.5	12.4
HungExpo (organizes trade fairs)	22.1	27.6	1.57
IBUSZ	28.6	125.7	15.2
IDEX (industrial trading co.)	58.7	29.3	1.19
Interglob Co. (transport co.)	14.2	19.0	.28
Kner Printing Co.	20.9	30.9	2.24
Kunep Co. (housing construction)	6.9	8.5	.16
MEH Scrap Processing Trust	50.3	118.1	11.0
Pannonia Hotel and Catering Co.	82.7	69.5	7.85
Pannoplast Co. (plastics manufacturer)	55.0	73.0	5.51
Pietra Building Ceramic Co.	13.0	13.2	.75
Richter Gedeon Chemicals PLC	233.1	207.7	12.5
Salgotarjan Plate Glass Factory	23.0	24.5	2.07
TRITEX Trading Co.	9.3	22.3	1.15
Volantefu Co. (transport company)	32.1	49.1	2.15

TABLE 4-2: AGGREGATE ECONOMIC STATUS OF 20 FIRST PRIVATIZATION PROGRAM COMPANIES AS OF DEC. 31, 1989

(Hungarian Accounting Standards)[351]

Performance Measure	Billions of HuF	Millions of USD
Total Equity	33	507
Total Assets	73	1123
Total Annual Sales	93	1431
Total Pretax Profit	6	92

3. Activate the government's privatization policies
4. Avoid conflicts in a smooth transition to a private economy
5. Improve the economy and competition
6. Expand the capital and stock markets
7. Generate proceeds for the government
8. Encourage small investors in Hungary to invest
9. Improve the profitability and efficiency of the companies involved[344]

The SPA succeeded, at best, in accomplishing the first three; the other six aims were clearly not met.

Twenty large companies, representing about 1 percent of all state assets,[345] were chosen from the fields of manufacturing, hotels and tourism, commerce, trade, and transport, and the list was made public on September 14, 1990.[346] Table 4-1 lists the twenty companies and presents some basic data concerning them.

Those chosen were selected on the basis of six criteria, a review of which makes it clear that the SPA was seeking first and foremost a successful, high profile, big splash program that would attract a lot of Western investors' attention to get itself off to a good start. Although some observers criticized this emphasis on the big, financially strong companies as backward,[348] the SPA looked for momentum.[349] The program attracted a lot of attention but fell far short of the desired effect.

The six criteria by which the 20 were selected reflected the emphasis of the SPA on large companies of high quality:

1. Suitability for quick privatization
2. Good recent performance

Privatization Under the State Property Agency

3. Support of the management and the employees
4. Suitability for a range of privatization techniques
5. Readiness for privatization
6. Potential investor demand[350]

As Table 4-2 shows, this was a group of large companies with decent earnings histories.

The transformation of all twenty into share companies was obviously the first step. After that, the SPA would rely on professional advisers to come up with a plan for privatizing an individual company. The process was to involve six steps.

First, details concerning the businesses and financial conditions of the twenty state companies would be disclosed by the SPA. Second, the SPA would choose one adviser to lead each company through the process. It would invite tenders for the positions of official advisers to the SPA for a specific company, although any firm seeking to become an adviser could tender for more than one position. Specific criteria were developed by the SPA concerning who would be eligible, and a firm submitting a tender would be required to submit a preliminary proposal outlining how it would go about privatizing that company. Based on the tenders, the SPA would choose an adviser for each company. In submitting their proposals, the would-be advisers were told that there were three possible privatization routes: a public offering and listing on the Budapest Stock Exchange[352] and perhaps other exchanges as well, competitive tender offers, ESOPs up to approximately 5 to 15 percent of the equity. The advisers were encouraged to be creative and were told that the three methods could be combined. The SPA suggested that some of the 20 companies might be ripe for breaking up into smaller companies before privatization. Advisers were expected to maximize prices by seeking multiple bidders, preferably inducing an auction.[353]

Any buyers brought into the process by an adviser would be subject to scrutiny for qualities the SPA deemed important. Strategic fit would be important in the manufacturing sector, with a preference given to bidders who could bring new technology, management techniques, and markets. Foreign participation was encouraged, subject only to national interest. Within Hungary, buyers representing smaller investors (e.g., pension funds) would be given priority. Individual investors in Hungary were told they could get loans from the Hungarian National Bank.[354] That nod to the individual investor was disingenuous; everyone understood even before the process was started that individual investors would not be participating in the FPP but would instead have to wait for the privatization of small shops and restaurants, which would be handled separately.[355] The FPP was from the start aimed almost exclusively at foreign investors, whose hard currency was coveted by the government.

Step three in the process came after an adviser was chosen. The company was then required to submit a more detailed privatization plan that would be subject to SPA approval. Fourth, the SPA would ensure that the company involved went through the necessary restructuring; in most cases this meant transformation into a share company. Fifth, the advisers would implement their plans and privatize the companies. Finally, the SPA would monitor each of the privatized companies for a period of two years.[356]

The SPA was clearly full of enthusiasm. It produced a slick brochure outlining the program and presenting basic information concerning each of the companies, with suggestions about the best ways to privatize. The SPA also advertised in that brochure more detailed "information packages" concerning each of the companies, for sale at between $50 and $200, depending on the company. It projected proceeds in the range of 25-40 billion HuF, or about $385–615 million, and predicted that the twenty firms' transactions would be completed by mid-1991.[357] The task was daunting; the SPA wanted to sell, in Hungary and abroad, a majority of the shares of each of the companies.[358] Yet the SPA was clearly anticipating multiple bids for each of the companies. It developed a complex evaluation scheme whereby only 25 percent weight was accorded to the price offered. The other 75 percent was broken down into four criteria with a total point value of 100.[359] The SPA would invite the firms with the three highest scores to compete in a final round.[360]

Not everyone was so optimistic. One German financier asked the question, "Just who is going to buy all the companies the ... Hungarians plan to sell?"[361] According to Merrill Lynch executive Adrian Friend, "Foreigners may be willing to take the family silver, but they're going to be far less interested in the base metal."[362]

The troubles began right away. The SPA took until December to choose advisers for the companies' privatizations, although only fifty applicants submitted tenders.[363] The staff of the SPA had not even come up with forms of agreements for the chosen advisers to sign, and it remained unclear where the money for paying the advisers would come from.³⁶⁴ Once the advisers were chosen, they had to untangle the complicated financial and organizational structures of the firms.³⁶⁵ More months slipped by.

By August of 1991, when the whole program had been scheduled for completion, SPA officials were publicly expressing their disappointment. Not one of the privatizations was complete by that time, and SPA director Peter Rajcsanyi attributed the pace to the "painstaking care" being taken by the advisers and an unfavorable international capital market.³⁶⁶ Others were critical of the whole process, arguing that spontaneous privatizations had worked much better.³⁶⁷ By September, there was little enough interest in seventeen of the twenty companies that they were being derided publicly as "wallflowers."³⁶⁸

A month later, however, SPA officials were telling reporters that although not a single transaction had been closed, the delays were really caused by the deluge of offers pouring in.³⁶⁹ This sort of statement is hard to square with what happened fewer than two and one-half months later: in January 1992, after a year and a half of operation, the First Privatization Program was scrapped.³⁷⁰ Although the SPA maintained it would continue to try to privatize the twenty firms on an ad hoc basis,³⁷¹ by mid-1992 still not one deal had been completed.³⁷² The complete failure of the SPA's first organized foray into privatization was hugely disappointing, particularly given the earlier furor over the IBUSZ transaction.

What went wrong? Certainly the disappointing results of this first program came about as a result of a combination of factors. Lajos Bokros, then president of the Budapest Stock Exchange and an SPA board member, tried to deflect the criticism as resulting from unduly optimistic expectations on the part of the public. "People think that privatization is just a matter of making an advertisement: this is for sale," he said. "Very few people know what is involved."³⁷³ Unfortunately, many of those people were working for the SPA. As the SPA deputy managing director, Karoly Szabo, conceded in mid-1992, more than a year after managing director Lajos Csepi had predicted the whole thing would be wound up, "The process is far more complicated than we ever thought."³⁷⁴ Every one of the companies needed substantial legal, financial, and organizational restructuring. The valuation process took far longer than anyone anticipated; the iconoclastic Hungarian accounting system, combined with the constantly and rapidly changing (mostly declining) financial picture of the firms, gave the auditors nightmares. According to one senior government official, "Until a few years ago, the only reason most companies kept accounts was to give work to another 250 people."[375]

In addition, the auditors had to go back many years, because the firms had not been undergoing audits on a regular basis, a practice labeled by Bokros "a bad mistake."[376] The advisers were swamped with inquiries from the West, but no one was buying at the prices the SPA wanted.

The SPA was also moving more slowly than it might have, in part due to the staffing and money shortages discussed earlier in this chapter, in part due to bureaucratic crawl,[377] and in part due to its insistence on multiple bids for companies and its unrealisticly high prices. The SPA had been stung once, in the IBUSZ public offering, by selling at what the public considered too low a price. It was determined not to make that mistake in the FPP. But while the SPA dawdled and negotiated, the companies it was trying to sell were sinking further and further into the worsening Hungarian economy.

On top of these pervasive problems, each of the twenty companies had idiosyncratic problems of its own. The story of the Danubius Hotel and Spa Company is illustrative.

Danubius was the flagship of the First Privatization Program. It was a company of high visibility in the West, owning the ultraplush Budapest Hilton, eight other large hotels throughout Hungary, and a dozen smaller hotels. It also had interests in a number of restaurants and some of the swankiest casinos east of Monte Carlo. Danubius was also highly profitable. In 1990 its pretax profits were up 70 percent over 1989 to approximately $12.8 million on sales of approximately $64 million.[378] In 1991 sales were up to $70 million, and profits went to $13 million.[379]

Delays over selection of an adviser ate up an inexplicable five months. The bid to do the privatization work was won by the London office of Nomura Investment Bank, which had hopes of privatizing the firm through simultaneous offerings on the Budapest and London stock exchanges.[380] But the frustrations came immediately. Local councils, the units of local government in Hungary, owned the land on which 18 of the hotels sat. They had contributed the use of that land to induce the building of the hotels but retained ownership of it. The first step toward privatization was transforming Danubius into a share company, which required the company to decide who would get how many shares. Each of the local councils involved demanded a percentage of the shares, and Nomura quickly found itself engaged in eighteen separate negotiations.[381] And this was only preparatory restructuring.

Danubius also ran into snags with the auditors who were trying to assess the accuracy of management's valuations. Those auditors found that many claimed assets were actually pledged as security for outstanding loans. They also found that many assets were in the form of potentially uncollectable accounts receivable.[382]

These negotiations and audits were still going on when, in October of 1991, two government policy changes ruined the transaction's viability. First, the government removed certain tax breaks that would have benefited the buyer. But the crushing blow came when the Hungarian National Bank announced that it would insist that any buyer pay market rates on subsidized loans Danubius had received years earlier.[383]

In the mid-1980s, the Hungarian National Bank borrowed approximately $300 million in hard currency from Austria at 8.5 percent and lent it to hotel companies (including Danubius) to finance construction. The loans to the hotel companies, though, were at 6 percent and in Hungarian forints. As the value of the forint declined relative to hard currencies, the Hungarian National Bank was losing on the exchange rate as well as the interest rate. In June 1991, as talk of privatization of the hotels was spreading, the Hungarian National Bank began renegotiating the loans to the hotel companies. It took the position that, in case of privatization, the hotels should either pay off the loans or accept a market rate of interest.[384]

In October, as the Danubius privatization was making progress, the Hungarian National Bank moved unilaterally, announcing that upon sale, the interest rate on the $24 million debt owed by Danubius would go from 6 percent to a market rate of 22 percent. The move was understandable in the political context of the time. If the subsidized rate were assumed by the borrower, the Hungarian National Bank would be in the sticky position of subsidizing some wealthy foreign corporation at a time of budgetary difficulties at home. Foreign dominance in the privatization processes was already a sensitive political subject, and the Hungarian National Bank and ruling MDF coalition may have decided that it would be too risky to allow the subsidized loans to be assumed. There would certainly have been criticism. That criticism might have jeopardized public support for privatization generally, or it might simply have lessened the likelihood that the ruling coalition would stay in power. Whatever the reason, with the bank's announcement, the Danubius deal was dead.[385] The SPA said it would try again in perhaps two or three years.[386]

That was a serious blow to the FPP. "This was the best prospect for an international public offering for any of the Hungarian companies in the next two years," lamented one SPA official.[387] "This is a serious failure in the history of Hungarian privatization; this would have been an important step," echoed another.[388]

It was also a blow to the value of Danubius. In June 1991, London-based experts estimated the value of the firm at $200-300 million. By February 1992, that estimate was down to $100-150 million. Deputy General Manager Imre Deak of Danubius pointed out that the drop was not attributable to declines in performance; Danubius was doing very well under difficult economic conditions. Rather, he pointed to the slow pace of privatization as the number one factor adversely affecting the value of his firm.[389] Seen one way, to avoid an 18 percent subsidy on a $24 million loan, the government and its central bank had allowed the value of one of its prime assets to decline by $50-100 million.

It was a blow not only to the SPA, the FPP, and Danubius, but also to the fledgling Budapest Stock Exchange, then little more than a year old and struggling with declining share prices, declining volume, and little interest on the part of companies in listing their securities there.[390] Nomura's plan had been to offer Danubius shares on the BSE.

In November 1991, after the collapse of the Danubius deal, the SPA's Peter Rajcsanyi said he doubted the BSE would be used for privatizations anymore. The market for IPOs was simply too weak.[391] These statements were made less than a year and a half after the IBUSZ offering was oversubscribed by 2,300 percent. The honeymoon was over.

Eventually, the idea of privatizing Danubius through an initial public offering on the BSE would be revived.[392] There would be numerous delays and postponements,[393] but by the end of 1992, Danubius' privatization had begun with a public offering of shares,[394] although not as part of the abandoned FPP, which was dead by the end of 1991. And in the early autumn of 1991, the future of the head of the SPA was again much in doubt.[395]

The Second Privatization Program: Stripped Companies

Although it was the First Privatization Program that garnered most of the headlines, it was not the only activity at the SPA in the first years of the 1990s. The Second Privatization Program (SPP) had a much different set of motivations, however, and a much different list of companies from the FPP.

The SPP was announced in March 1991, at a time when the FPP was far behind schedule and already showing signs of unraveling. It was in part investigative, in part punitive, and only in part aimed at furthering the goal of integrating state owned assets into a growing private economy. The SPP involved the privatization of those companies whose assets had been stripped during the frenzied days of unsupervised spontaneous privatizations.[396] According to Lajos Csepi, head of the SPA, the goal of the SPA in the Second Privatization Program was primarily investigative: the SPA wanted to see clearly what happened to the assets of certain spontaneously privatized companies when they were transformed into joint stock companies. He hinted broadly in September 1991 that the government was going to be looking into what it suspected was looting on the part of the former Communist nomenclatura.[397]

The concept behind the SPP was the privatization of those "shell" companies, still owned by the state, that existed after the companies either contributed all their assets to another economic association (e.g., a joint venture with a foreign partner) as a noncash contribution or leased out some or all of their property. The state was left holding these empty shells, and in most cases all that was left were the undesirable assets some acquirer did not want. There should have been valuable interests in whatever economic association to which the assets had been transferred, but as noted above,[398] such was not always the case. The SPA feared that the assets left in the shells would continue to disintegrate through neglect or malfeasance. It wanted to use the privatization process as an opportunity to find out where (and to whom) the assets had gone and what was left.[399]

When it announced the SPP, the State Property Agency listed four aims of the program:

1. Reduction of state ownership to develop a market economy
2. The tracing and defining of status of state property

3. Generation of income to reduce state debts
4. The establishment of profitability and orderly liquidation[400]

The only criterion for inclusion in the SPP was that the company have put at least 50 percent of its assets into some other economic association or have leased out at least 50 percent of its assets. The initial plan was to publish a list of all such companies, group them into groups of 15, and privatize one group every month.[401] In practice, though, after the first group of 12 was announced, the SPP was simply absorbed into the uncategorized general privatization. The program went forward but not as a separately organized process.

The SPA clearly anticipated some hostility to its investigatory privatization. In announcing the program, the SPA noted that the cooperation of management was "welcomed but not required." It made clear that the SPA would decide who, when, and how to privatize these firms.[402]

Table 4–3 presents basic economic data for the companies included in the first phase of the SPP. Several facts are of interest in looking at the table. First, the companies were exclusively industrial manufacturing firms that were small to midsize in terms of sales. But they were employers of large numbers of workers. This is interesting for a couple reasons. First, if there was any looting going on by the nomenclatura, there were tens of thousands of individuals whose livelihoods were being jeopardized. Second, these firms were grossly overstaffed. They had cumulative sales of approximately $252.3 million (at 75 HuF per dollar) but more than 23,000 employees. That works out to only approximately $10,740 in sales per employee. These were more jobs centers than industrial facilities.

Although this initial list included only twelve companies, it was thought of all along as only the first phase, and the managing director of the SPA told an interviewer later in 1991 that he thought there were 60 to 70 companies that came under the SPP criteria.

By the time of the SPP's inception, the State Property Agency had learned its lesson regarding brave optimistic forecasts of rapid completion for its privatization programs. As a result, it was never necessary to scuttle the SPP even though it moved very slowly. The SPA was able to find buyers for some of the companies, it was able to find buyers for parts of others, and still others languished without buyers at all through 1992. The SPP was more an investigation and orderly liquidation than a serious attempt to sell going concerns.

TABLE 4-3: COMPANIES OF THE SECOND
PRIVATIZATION PROGRAM, FIRST PHASE[403]

Company	Sales	Pretax Profit	Employees
Machine Tool Industrial Works	1780	(110)	3610
Szatmar Furniture Factory	968	13	887
Information Technical Co.	860	18	502
Screw Industrial Co.	1288	58	2268
Construction Machine Industrial Co.	407	(101)	649
Food Industrial Machine Co.	2631	(150)	2657
Elegant Majus 1 Garment Factory	1175	15	2356
Ganz Machine Tool Factory	151	9	208
Hungarian Optical Works	2651	(83)	3738
Grabocenter	5273	302	2837
Csepel Works' Garment Industrial Machinery & Bicycle Producing Factory	1658	(108)	1977
Chair & Upholstery Industrial Co.	184	(10)	1802
Total	18,926	(147)	23,491

The Third and Fourth Privatization Programs: Targeted Industries

The third and fourth privatization programs, announced with little fanfare in April 1991, were industry-specific. The third was directed at the construction industry, and the fourth was directed at Hungary's historic wine-growing regions, particularly Tokay, where some of the world's greatest sweet white dessert wines have been produced for centuries. Both were crisis situations that needed quick attention.[404] The construction industry was falling apart and was thought to be uniquely important because it would be a necessary piece of infrastructure in rebuilding other sectors of the economy. The wine companies were unique because the cultivation of excellent mature vines takes years. If they are allowed to fall into an unhealthy condition, they may easily become diseased and unfit to produce the quality of wine Hungary was known for. If the vines had to be replanted, it could be twenty years before they were producing the same quality grapes as before. It was therefore a matter of national pride that that industry not be allowed to fall.[405]

The third and fourth privatization programs were neither more nor less successful than privatization generally. Assets and companies were sold, but everything moved more slowly than many would have liked.[406]

The Small Retail Shop and Restaurant Privatization

There simply was no way to avoid foreign dominance of the larger privatization programs; individual investors had neither the will nor the money to compete with foreign bidders for large firms, and the capital markets were not sufficiently developed to permit them to raise the money or pool their money into investment funds. From the earliest days of privatization, therefore, the Hungarian government envisioned a separate program for the sale of 8,000 to 10,000 small shops and restaurants.[407] It was hoped that this domestic participation would not only enhance the peoples' stake in and popular support for economic and democratic reforms, but also begin to create the middle class that many saw as the engine that would drive the country's future economy.[408]

The necessary legislation[409] became effective in early 1991. It had been controversial in Parliament the year before, and the arguments over it were representative of the debate over the pace of privatization generally. The legislation called for an auction process with minimum prices carefully controlled by the SPA and not open to foreign participation. The opposition parties wanted the process speeded up, with more emphasis placed on the establishment of a private economy than on the generation of revenue and the participation of domestic investors.[410] The ruling coalition held, however, and the more controlled process was legislated into place.

By most measures, the program was a bust.[411] The process began in April 1991. There were 10,240 enterprises in the program, but only 3,211 ever came up for auction.[412] In the early stages, most businesses failed to generate even a single bid. By October 1991 only approximately 100 were sold.[413] After that, the SPA began lowering its minimum bids to more realistic levels. In the end, however, only 1,734 (or 16.9 percent of those in the program) were sold.[414] Why?

As in the First Privatization Program, the failures here came about as a result of the concatenation of numerous factors, most of which were beyond the control of the SPA. First, after the program began and SPA officials began looking carefully at the businesses to be auctioned, they discovered a mess of unclear property rights in many of the businesses. Approximately 4,000 of them were subject to long-term leases and therefore could not be sold. In others the state's ownership was unclear enough that it would offer only leaseholds for sale in the auctions, which scared off nearly all possible bidders.[415]

Second, as in the FPP, the State Property Agency set the prices it was willing to accept at too high a level.[416] One cannot be serious about creating a system of market prices unless one is willing to accept the fact that the market will sometimes set a low price. But the prices set by the SPA as minimums were too high only because of the most damaging problem facing the privatization of small shops and restaurants: lack of credit. The lack of money available to lend to would-be buyers effectively made the sale "cash and carry." We should not be surprised, then, that most Hungarians, whose average wage was a little more than $200 per month (at least by official measurements),[417] could not afford the prices set by the SPA. Observers of the difficulties encountered by the SPA in auctioning off these small businesses mention repeatedly the role played by the paucity of credit.[418]

The banking system in Hungary, notwithstanding its reform in 1987, still works like a central state bank. The commercial banks do not behave like commercial banks, and they certainly do not occupy the role of a small business lifeline the way they do in other market economies.[419] Small business owners could not simply walk in and get loans the way they can in many Western countries. One of the main reasons is that when the commercial banks were spun off from the Hungarian National Bank in 1987, each was given a portfolio of loans and therefore an established clientele. Those clients were (like the banks) state-owned companies for the most part and were dependent upon continued bank loans for survival. For a commercial bank to refuse further lending could easily cause the client to default on existing loans. Given the thin capitalization of the commercial banks and the fact that most of that capital is in the form of notes payable from its state-owned clients, that would be a dangerous policy in which to engage regularly. In addition, the relationship with its existing clientele was a cozy one, and the lending was easy. So the commercial banks kept lending their available funds to the large, moribund state-owned enterprises and neither needed nor wanted the business of more entrepreneurial clients. Making the problem worse was the inflation rate, which was in the mid–30s in the late 1980s and came down to the mid-20s by the end of 1992. With inflation high, the Hungarian National Bank had to keep a tight money policy in place,[420] thus driving up interest rates and drying up credit.[421] Going to the bank to borrow money with which to buy a business being auctioned was simply not an option for most Hungarians.

Illustrative of both the success and failure of the program was the auctioning of the Veszprem Catering Company. The Veszprem Catering Company was started in the 1950s to provide the Lake Balaton region of Hungary with food service operations. Lake Balaton, in southwestern Hungary, is the largest lake in Europe and is a popular tourist and vacation spot for Hungarians and foreigners, particularly Germans from both Germany and Austria.[422] In the spring of 1991, ten of Veszprem's constituent businesses were being auctioned by the SPA. Only three had bidders at all. Two of those were able to finance their purchases with a combination of bank loans and substantial savings.

The first was the Cosy Nook Cafe and the attached Cubby Bear Bistro in Balatonfured. Jozsef Turi, a 55-year-old restaurant manager won the bidding with a bid of approximately $270,000, a sum representing approximately ten years worth of wages for the average Hungarian. His family had started its own restaurant in 1970, and accumulated savings from that venture made up his down payment. He was able to get a bank loan for the balance: 20.5 percent fully amortized for a six-year term. "I want to create a high quality restaurant, so we will have to renovate it. I am toying with the idea of talking with McDonald's about coming in," stated Turi enthusiastically after his winning bid was confirmed. Ildiko Keskeny, the lawyer supervising the auction on behalf of the SPA was beaming. "It was very exciting."

Equally enthusiastic were the winning bidders for the Wild

Duck, a squat, white-washed restaurant and discotheque on the banks of Lake Balaton. Ervin Hartai and his brother won the auction simply by placing a bid for the minimum: about $81,000. They were able to scrape up their down payment from savings and then borrow the rest from a bank at 20 percent over six years. They planned to use other savings to update and enlarge the Wild Duck. They were optimistic about their chances of success on the morning of the auction. "It didn't used to be a popular spot, but it will be now," said Hartai.[423]

Unfortunately for both the SPA and potential entrepreneurs in Hungary, most Hungarians had neither substantial savings to make down payments nor the ability to secure bank loans no matter what the interest rate. Hence the lack of success in the program.

Clearly 1990 and 1991 were not good years for the State Property Agency. It came in planning massive privatizations at both ends of the spectrum. The First Privatization Program emphasized crown jewels at lofty prices and targeted the foreign markets. This was to ensure international momentum for the privatization programs. After IBUSZ, there was reason to be optimistic; foreigners seemed to be drooling for the opportunity to buy Hungarian businesses. At the other end of the spectrum were the auctions of small retail shops and restaurants. Those sales of small businesses were to be targeted exclusively at the domestic Hungarians who might otherwise have resented the foreign dominance of the privatization programs. In addition, sales of more than 10,000 small businesses would go a long way toward cementing a bourgeoisie that would then support democratic and market reforms and build the economy. Good strategy never succeeds without sound execution, however, and the SPA did not execute soundly.

But the Hungarian people have a reputation for being by nature pragmatic, creative, and resilient, and it is a testimony to their pragmatism that rather than giving up on privatization altogether, the SPA kept trying to sell businesses (or parts of them) and coming up with new solutions to problems it was encountering.

Five

Changing Course in Privatization — More Programs with Greater Flexibility

In 1991 and early 1992, the SPA tried a number of alternative methods of privatization, searching for programs that would work. It came up with no fewer than seven during that period, and although those programs were various, the SPA never deviated from the basic principles that underlay, for better or worse, the MDF-led coalition's plans for privatization: sales rather than give-aways, insistence on what it considered "fair prices," and reliance on foreign participation while always remaining on the lookout for ways to facilitate domestic purchases.

Investor-Initiated Privatization

As early as the fall of 1990, the SPA announced itself to be receptive to offers by potential investors for specific companies they were interested in acquiring. But this was at a time when the SPA still did not realize the complications involved in privatizing state-owned enterprises. The SPA at that time promised to respond to any preliminary proposal by an investor within 30 days. Then, assuming a favorable SPA response, the investor would have 30 to 90 days in which to formulate a final offer. Within that 30- to 90-day period, promised the SPA, the company would be audited and converted to a joint stock company. If no competing bids appeared and if the bid was in line with the audit, the SPA would accept the bid. SPA evaluation was expected to take a scant three weeks.[424] In retrospect, these time tables seem almost comically optimistic, particularly since at the time they were made the SPA staff numbered only about 40.[425]

But the SPA did, in early 1991, come up with a more realistic formal program for investor-initiated privatizations, raising the possibility for the first time that hostile takeovers would be possible.[426] It was looking for ways to speed up the process of privatization and saw investor-initiated deals as one way to do that.[427] Indeed, in the publication distributed in the spring of 1991 announcing the program, the SPA made it clear that speeding up privatization was the goal of this mechanism.[428] This made the fourth method of privatization, the other three being SPA-initiated privatizations (as in the First through Fourth Privatization Programs), spontaneous privatizations, and the auctions of small retail shops and restaurants.[429] Offers for state assets could thereafter come literally from anyone; even employees and managers could submit bids. Indeed, the SPA made it clear that it would be willing to consider favorably lower offers from employees,[430] and there were favorable terms available.[431]

Any company was available for an investor overture except those falling in one of six categories:

1. Those small shops and restaurants being auctioned only to Hungarians
2. Firms already being privatized at SPA initiative
3. Firms that had already been through the privatization process
4. Public utilities
5. Financial institutions (banks were a hot commodity and would be separately privatized)
6. Companies that held more than 40 percent of the domestic market in their area of main activity[432]

The SPA published with its announcement specific guidelines for submitting offers and made it clear that once it received an offer, it would seek competitive bids for a period of 30 days.[433] This seeking of competitive bids was a real problem. The government considered it a matter of principle that it receive the highest possible price for its assets, and toward that end it insisted in all its programs on some sort of competition. The difficulties entailed generally by that policy are explored fully below.[434] But if the goal of investor-initiated privatization was to speed up the process, this policy obviously ran counter, even though the SPA tried to ameliorate that problem by setting a 30-day period.

Delay was a problem inherent in the policy of competitive bidding generally and not specific to the investor-initiated program. But the policy was particularly destructive in the investor-initiated program. No potential bidder wanted to sink large sums of money investigating and identifying a target only to end up a stalking horse for another rival bidder that could free-ride on its sunk investigative costs. The unmeasurable but undeniable effect of such a policy was to discourage all bidders from doing the investigation that might lead to attractive bids. By demanding the right to seek competitive bids, the SPA was shooting itself in the foot, probably decreasing rather than increasing the revenues from this program.

Although the investor-initiated privatization program opened the door for hostile takeovers, they occurred rarely. The full-blown battle for corporate control, in which a defensive management uses the means at its disposal to keep a hostile acquirer from purchasing shares from the stockholders, never really materialized. The management position was always too weak for it to put up a fight: the defensive tactics at its disposal were few, and its shareholder (the SPA) was always inclined to sell at the right price. All management could really do was either delay the process, hoping thereby to wear out the foreign bidder, or attempt to make the company attractive enough to those valuing the company on behalf of the SPA that the price would be set unrealistically high. Most managers opted instead to try to make *themselves* attractive to the foreign bidder in an attempt to cut a postprivatization employment deal.[435] And, in fairness to the managers, many (perhaps most) were eager for privatization and an opportunity to manage their companies without undue bureaucratic control.

The Asset Management Program

The statute authorizing SPA activity allowed the SPA to sell assets or lease out their management.[436] Thus, the SPA held open the possibility that potential investors could not only bid for the purchase of state-owned assets in the investor-initiated privatization program, but could also bid solely for the right to manage state assets.[437] In early 1991, the SPA published a brochure pointing out that it could lease out state property for management while retaining ownership. The brochure which described various ways in which interested investors might participate, was comprehensive and wide-ranging, but also general and lacking in specifics. It was a statement of general principles that would apply in any future asset management program. The SPA did, however, develop later that year a more formalized asset management program that held great promise but went nowhere the first time it was tried.

The program, announced in April 1991,[438] was a testimony to the creativity of the policymakers at the SPA. The SPA had chosen five companies for the experiment: Carbon Rt., Pecsi Agroker Rt., EGUT Rt., Pecs Construction Kft., and ARBAV Construction Limited Liability Co. The five had two things in common. First, they had already been transformed into share companies, and second, each desperately needed new management but was unlikely to generate enough investor interest for a sale at a price commensurate with what the SPA considered its potential. The five companies were also concentrated in the construction industry, which had been targeted for early privatization under the Third Privatization Program discussed earlier in this chapter.

The program called for an interested party to bid for the right to manage a percentage of the assets of the companies, ranging from a minimum of 10 percent to a maximum of 76.68 percent, for three years. Those assets would then be transferred to the manager, who would own them outright, subject only to a contractual obligation to remit a certain sum to the SPA at the expiration of the contract under the following terms:

1. The manager had to return to the SPA the value of the assets, measured at the time of initial transfer. The assets were not to be returned to the SPA, however. Rather, the SPA was to get at least 70 percent in cash and the rest in marketable securities.

2. If at the end of three years, the increase in the value of the assets over their value at the time of initial transfer was less than 10 percent, the SPA would get 100 percent of the gain. The profit, like the principal, was to be paid in cash and marketable securities on the 70/30 formula.

3. If the increase in asset value was between 10 and 25 percent, the SPA would get 70 percent of the gain.

4. If the increase in asset value was greater than 25 percent, that increase would be split evenly.

5. Any dividends on investments made by the manager with managed funds, up to 50 percent of net profits, were to be split 50/50.[439]

Given the amount of work that obviously went into the development of the program, it was a shame that it flopped, but it did. The SPA received only three offers, and it was able to conclude a deal with none of the offerees. The two biggest problems were that it was unclear how inflation and currency exchange fluctuations would be accounted for, and from the point of view of the potential manager, the risk/reward ratio was simply too high.[440] The managers were, in effect, buying the assets with payment delayed three years. But they were asked to take all the downside risk and share the upside potential with the SPA 50/50 at minimum. But, with a note of cheery determination, the SPA told reporters in June 1991 that the first asset-handling tender had simply been an experiment and that it would try again later.[441]

The SPA did try again, with some success. In May 1992, the Hungarian brokerage firm Co-Nexus won out over Daiwa Securities in the Second Asset Management Tender by promising to pay cash, instead of securities, to the SPA at the end of the contract period. Under the contract, Co-Nexus took over the SPA's ownership position in eight companies with a nominal value of 3.953 billion HuF (approximately $53 million) and a market value of about 60 to 70 percent of that figure. It took full ownership rights for five years through a subsidiary set up for that purpose. Co-Nexus was required under the contract to file annual statements, and if the value of the portfolio declined, the SPA was authorized to terminate the agreement and seek damages against Co-Nexus. But, unlike the First Asset Management Tender, the second one permitted Co-Nexus to keep any profits after paying back the principal to the SPA at the end of the five-year period of the contract.[442] This was the first formalized program (as opposed to ad hoc sales of state-owned enterprises) in the SPA's privatization efforts that could be deemed a success.

Self-Privatization:
Spontaneous Privatization Redux

The most successful of the SPA's privatization efforts came, ironically, through SPA-controlled spontaneous privatizations. Recall that it was the scandal surrounding various pre-SPA spontaeous privatizations that led to the creation of the SPA. After attempting to push certain companies into formalized programs such as the First Privatization Program, the SPA found that it acted best when it acted least. Put another way, the SPA discovered that investors and company managers are better identifiers of candidates and initiators of privatization than is a state agency.

By the summer of 1991, the privatization process was bogging down, and the SPA was coming under fire for its several failures and its inability to come close to early predictions about the pace of the process. In particular, there was concern that bureaucratic delays were chasing buyers away and resulting in the diminution of companies' values as the economy slipped further. According to one Western accountant, "The delay has meant these state companies have carried on bleeding to death."[443]

The frustrations of the Colgate-Palmolive company are typical. In 1991, Colgate was interested in acquiring the large Caola consumer products firm and entered into negotiations with the SPA for its purchase. The auditors valued the company at between 3 and 4 billion HuF, but Colgate was willing to pay only 2 billion. The Americans eventually became so frustrated with the protracted negotiations and the SPA's intransigence on the asking price that they walked away.[444] The result was that by the end of 1992 the SPA was still trying to sell Caola, but at a much lower price. This company's former dominant market share in detergents and toiletries had slipped drastically (to approximately 30 percent in 1992), and production was cut 25 percent in 1991 alone.[445]

Managers at Vegytek, the Hungarian chemicals trading company, experienced similar frustration in one of the first transactions reviewed by the SPA. EIA Acquisition, an American firm, had offered $10 million for a majority stake in Vegytek, but the SPA blocked the deal in May 1990, saying that the price offered was too low for a company with sales (under the command economy with its fixed prices) of $7 million per year.[446] Although the Vegytek officers pleaded that they were in desperate need of cash, and although there were no better offers being made, the SPA refused to relent.[447] "It's obvious the bureaucracy is a bottleneck," confessed SPA chief Lajos Csepi.[448]

On June 2, 1991, the Ownership and Privatization Committee of the government's economic cabinet submitted to Parliament its draft policy for privatization efforts. Among the draft's many recommendations was that the privatization process be decentralized and accelerated by implementation of a simplified process for small and medium-sized firms.[449] Two weeks later, the Finance Ministry, which was not happy with the committee recommendations, issued its own proposed policy on privatization. In announcing his recommendations, the Finance Minister agreed with the committee that the privatization process should be decentralized but suggested that the SPA's input should be curtailed and that more self-privatizations be permitted.[450]

There were signs in the summer of 1991 that the SPA's inflexible attitude on price was beginning to change as a result of the extensive criticism. One deal's success has been attributed directly to the complaints surrounding the SPA's attitude in the Colgate/Caola negotiations. Shortly after Colgate walked away from Caola, the Swedish firm Electrolux reached an agreement with Lehel, the Hungarian white goods manufacturer, for a 100 percent buyout. When the two firms presented the deal to the SPA, the SPA approved almost immediately and at the price suggested by the parties. By late 1992, the combination occupied one of the most stylish appliance showrooms in central Budapest's shopping district.

This was the climate in July 1991 when the SPA announced the first phase of its Self-Privatization Program. Companies eligible would be small: fewer than 300 million HuF (approximately $4 million) in sales, fewer than 300 million HuF in assets, and fewer than 300 employees. The companies meeting this description (thought initially to number approximately 300) were to begin the process themselves by transforming into share companies and choosing a privatization adviser from an SPA-approved list. Any firm that did not begin the process by the end of March 1992 would be forcibly privatized by the SPA.

Once a company chose an adviser, those two would handle the details of the privatization themselves, and the SPA's role would be limited to approving or disapproving the deal finally struck.[451] The SPA's only limitation on price was that it would accept not less than 80 percent of the value set by the auditors.[452]

In early July, the SPA asked potential advisers to apply by July 15 to be placed on the list of 40 approved advisers. They were told that they would be paid a flat 5 percent of the sales price with a bonus of up to an additional 5 percent if the transaction was completed by the end of 1991.[453] In addition, the advisers were told they could strike side deals with the companies they were advising whereby they would be paid for additional postprivatization consultancy services such as business plans and reorganizations.[454]

These generous terms attracted a lot of interest, with more than 200 firms meeting the deadline,[455] although many of the larger firms sniffed that the companies to be privatized were too small to waste their time on.[456] The SPA immediately announced that it was considering expanding the list of approved advisers to more than 40. The SPA also raised its estimate of the number of companies eligible for privatization under the 300/300/300 criteria to 350. The program would begin on September 1, 1991, and be completed by March 31, 1992, according to the SPA.[457] At the same time, the SPA announced plans to begin the same process later in 1992 for slightly larger firms.[458] Curiously, and perhaps unrealistically, an SPA official said in August 1991 that the SPA expected most investors to be managers and employees.[459] When SPA officials began taking trips abroad to generate foreign investor interest,[460] it became clear who the real potential buyers were.

Observers were optimistic, but wary. According to one American lawyer, "If the SPA really delegated the decision making, we can do deals in three months instead of a year."[461] The program was well received in foreign capitals,[462] and the European Bank for Reconstruction and Development announced that it was considering setting up a fund to invest in the firms being privatized in the program.[463] Most analysts, however, thought the companies in this first phase of the Self-Privatization Program were too small to make much of an impact on the structure of the economy.[464] The companies' aggregate book value was only approximately $333 million.[465] The real value of the program was obviously as a pattern for the future.

Unfortunately, the speed of the program looked more like a replica of the past. As of January 1992, only five of the almost 400 eligible companies had been sold, and only another twenty had completed the process of transforming themselves into share companies. Another 200 were said to be in the "preliminary phase" of privatization.[466] By the end of March, when the process was supposed to have been wound up completely, only five sales had been made.[467] By May 1992, of the 359 SOEs that had agreed to be privatized under the program, only 73 had been transformed into share companies, and only ten of those had actually been sold.[468] The process continued steadily but slowly thoughout 1992.

The disappointing pace of this decentralized system illustrated that intrusive bureaucracy at the SPA could not be blamed entirely for the slowness of the privatization process. As we will see later in this chapter, many factors combined to stifle the process, only some of which can be blamed on the State Property Agency.

And because the factors that slowed down the Self-Privatization Program were systemic and therefore unrelated to any particular method of privatization, there was a commitment on the part of authorities to stick with it. It may have been a slow process, but it was perceived to be the best route to privatization,[469] and the government made it clear that self-privatization would be the main vehicle for privatization in the future.[470]

Thus, in March 1992, as the 300/300/300 Self-Privatization Program crept along, the SPA announced a new, expanded program. This "second phase" of the program would involve slightly larger companies. They would have fewer than 1,000 employees and less than 1,000 million HuF in sales and assets. Their aggregate book value was estimated at between $933 million and $1 billion,[471] and 278 companies meeting the criteria appeared on a draft list being circulated in the spring of 1992.[472] By the end of 1992, progress on this program was moving along, again slowly. Standing together, the first and second phases of this Self-Privatization Program would not privatize the economy; by one estimate, together they represented at most 15 percent of GDP and their impact on the employment figures were not dramatic.[473] But they were a part of the solution. And the SPA had learned that, where possible, a more decentralized process, with most decisions being made by the economic actors involved, was preferable to substantial bureaucratic control.

Privatization Through Liquidation

Unfortunately, not all privatizations could be initiated outside the SPA, and not all state-owned enterprises were as attractive as others. Some companies were hopeless by the time privatization became a reality, and the only possibility was a salvage operation whereby assets were sold for whatever the state could get for them.

Even before the SPA made conscious decisions to privatize the assets of some companies through liquidation, however, a form of "spontaneous liquidation" was underway. In late 1991, it became apparent that many state-owned enterprises were slowly liquidating themselves on an informal basis by meeting current obligations through the sale of assets to private parties.[474] This created the illusion of financial survival at those enterprises when in fact a liquidation was taking place. The bad news for the state was that, as in the liquidation of any insolvent firm, the shareholder (the state) received no proceeds from the sales; rather, creditors were taking all. The good news was that at least the assets were being privatized.

The SPA often found liquidation the best, and at times the only viable, method of privatization. A good example was the liquidation of HAFE in 1992. HAFE's operations consisted of a factory that made painting equipment. In the late 1980s, approximately 50 percent of its output was exported to the Soviet Union. As the Soviet Union's economy and political structure collapsed, HAFE's fortunes turned sour. In 1988 it began losing money. By 1991, shipments to the former Soviet Union ceased altogether. In the first two months of 1992, orders were down 70 to 80 percent compared with the previous years. The company's markets had collapsed, and its business had ground to a halt. With no buyers for the firm in sight, the SPA in 1992 began selling off assets of HAFE for whatever it could get.[475] With assets worth approximately 2.4 billion HuF and debts of approximately 1 billion HuF, the firm was not insolvent, and the SPA could expect a return even after satisfying creditors.[476] The example demonstrates that SOEs may be solvent even though their businesses are completely irrelevant to the new market economy.

Hungary in the early 1990s was also engaged in the formal liquidation of insolvent state companies, privatizing the assets bit by bit. The State Liquidation Agency oversaw the orderly liquidation of moribund state firms until January 1992. In that month, the SLA became a share company, Reorg PLC.[477] Fifty-one percent of its stock was held by the Ministry of Finance, and the other 49 percent was taken by a consortium of four law firms and financial consultancies. At the time of its formation, Reorg PLC was handling approximately 150 liquidations, a number that would increase over the next few months as the state economy slipped further.[478] The company was established as a profit-making enterprise: under the new bankruptcy law,[479] the company handling a liquidation was entitled to a fee of 2 percent of sales.

Ironically, attempts by the state to realize some value out of SOEs by liquidating them may have impeded the privatizations of state companies as going concerns. By mid-1992, potential buyers of state assets had discovered an important point. Whereas the SPA was under pressure from the right (including the ruling MDF coalition) to hold out for high prices for companies being privatized, companies being liquidated were having their assets sold for whatever the market would bear. Thus, liquidation was seen as the "wave of the future," since assets could be snatched up much more cheaply if the buyer only wanted assets and was just patient enough to wait for the company to slip into a desperate state.[480] Potential buyers from the SPA therefore often had an incentive not to bid but rather to wait.

The Breakup and Privatization of Hollow Parent Companies

In April 1992, the SPA announced a new program to break up and privatize fourteen companies in which a parent corporation was left a hollow holding company. These companies were firms that had, through a series of transformations, created subsidiaries to which assets of the original enterprise had been transferred. The original state enterprise was then left in the position of a shell company holding nothing but interests in the subsidiaries and having the SPA as its only shareholder. Most had arrived at that condition through spontaneous privatization, meaning that the parent corporation's interest in the subsidiaries was that of partial owner (often minority owner), with a foreign company as joint venturer. Through the program announced in April 1992, the SPA sought to privatize its partial indirect ownership positions in the subsidiaries.[481]

The Asset Leasing Program

Recognizing the difficulty for many domestic Hungarians to come up with sufficient cash to bid against foreigners, the SPA set up an asset leasing program in November 1992. The program was made possible by the enactment that summer of Law LIV of 1992 on the Sale, Utilization, and Protection of Assets Temporarily Owned by the State.[482] In late November, the SPA published a list of eight companies that were available for lease. Under the terms of the program, the lessee would pay nothing at the outset. Rather, the lease payment negotiated would be payable in six or eight years (depending on the size of the transaction), or ten years in the case of agricultural companies. The only down payment required would be a financial guarantee in the amount of the first year's lease payment.[483] The idea was that a private operator could manage a state company more profitably than the state, since any gains in excess of the lease payment would belong solely to the lessee. The lease concept's advantage was thought to be in its allowance for control by managers unable to come up with a purchase price for the company. Indeed, only natural-born persons are permitted as lessees.

The first successful leasing transaction was not completed until early 1993, when a confectionary manufacturer was leased by the SPA on a long-term basis with the hope that the lessee would purchase the company outright at the end of the lease term.[484] Consistent with the motivation for the leasing plan generally, sale of the company through normal privatization channels was unsuccessful.[485] There were, however, fifteen bidders in the lease program, indicating that the program would be promising in the future.[486]

Six

The Banking and Agricultural Sectors — Special Challenges in Reform and Privatization

Although in one sense every privatization and instance of economic reform in Hungary was unique, two industries in particular presented special problems for the country. Those two industries were banking and farming, and they are taken up in this chapter. The variety of problems encountered and issues raised by the reform and privatization of these two industries illustrates just how complex was the task of transforming and privatizing an entire well-developed economy.

Reform of the Hungarian Banking Industry

Making the transition from a monolithic state controlled banking system to a system of competing private banks was one of the most daunting tasks facing Hungarian reformers in the late 1980s. Yet it also was one of the most important. A competitive market economy would require Western-style credit facilities. In Hungary, progress toward a system of private commericial credit and savings banks began in 1987. After a series of successful structural reforms, the country moved toward privatization of its banks. But the process became bogged down in economic emergencies and scandals involving criminal allegations.

Banking System Reform, 1987-1992

Preparation for privatization of Hungary's banking industry began with the implementation of regulatory reform in 1987. Until that time banking in Hungary was the exclusive province of the Hungarian National Bank. It served not only traditional central banking functions but also commercial banking functions, providing loans to state-owned enterprises.[487] As part of a general trend toward greater market orientation, the Hungarian Socialist Workers party in 1987 spun off most banking functions from the HNB into other institutions. The HNB retained the central banking functions (e.g., foreign exchange and domestic currency control), but spun off the commercial and savings bank functions to a number of separate institutions.[488] The purpose was to make credit allocation more efficient. In practical terms, that meant greater pressure on the state companies with outstanding loans, since the new commercial banks to whom their loans had been transferred were more likely to push them into bankruptcy than was the HNB.[489]

Thus in the sense that refusal to continue to fund inefficient enterprises represented more efficient allocation of Hungary's limited credit resources, the breakup of the banking system was a positive step. Perhaps more importantly, however, it paved the way for privatization of the banks that had been spun off. It did so in two ways: by creating independent business companies that could be transformed and sold, and by exposing the precarious financial position of those banks and thereby making privatization necessary.

The precarious financial position of the banks was a direct result of the HNB's spinning off both the good and bad aspects of its commercial banking function. The large banks had not developed in a market economy and therefore suffered from a lack of market discipline. In addition, they had simply inherited the clients and bad loans previously held by the HNB.[490] Each new bank was given capital and a portfolio of existing loans. Those portfolios were made up of loans the HNB had made to state-owned companies, and they were distributed roughly by industry. For example, the Hungarian Credit Bank received as clients 40 percent of heavy industry. The cooperatives, state farms, and tourist operations were given to the Commercial and Credit Bank. Budapest Bank was given the collieries.[491] Together the three banks held 70 percent of Hungary's commercial banking assets.[492] Each of the mentioned industries was

TABLE 6-1: DIRECT AND INDIRECT
STATE OWNERSHIP OF LARGEST BANKS
IN HUNGARY IN 1992 BY PERCENTAGE[497]

Bank	% Directly Owned by the State	% Indirectly Owned by the State[498]
OTP	100	0
Hungarian Foreign Trade Bank	44.6	42.9
Hungarian Credit Bank	49.3	50.6
Hungarian Commercial and Credit Bank	34.1	10.1
Budapest Bank	52.0	36.0

in terrible financial shape in the late 1980s and getting worse. In all, those three banks were allocated $670 million in bad loans, but by late 1991 their aggregate capital was only $480 million. Still, they claimed a profit margin of more than 60 percent and paid dividends of 12 to 15 percent annually. But the apparent prosperity was illusory: auditing standards were very loose, and accounting principles did not require writing off bad loans as would be the policy under Western principles.[493] Until new accounting standards were put in place in 1991, there was no accepted concept of the nonperforming loan.[494] In addition to the bad loans that were provided as the banks' "assets," each bank was allocated a portion of the debt incurred by the HNB to finance commercial ventures in Hungary. That debt totaled approximately $267 million in all.[495]

And although they were spun off from the HNB, the banks still suffered from their state ownership. For example, the largest savings bank, OTP, was 100 percent owned by the state.[496] In addition, as Table 6-1 shows, each of the five largest banks was owned substantially, either directly or indirectly, by the state as late as August 1992. Indeed, in 1992 approximately half of the nearly three dozen banks in Hungary were wholly owned by the state.[499] In all, the state directly owned approximately 35 to 40 percent of all bank shares, and the rest were owned primarily by state-owned enterprises.[500]

As a result of their central bank lineage and continued state ownership, the banks continued not to serve as commercial banks do in Western countries and to do poorly economically. Those two results are linked: the commercial banks continued to be tied to the existing clientele of state-owned enterprises that were in difficult financial straits. They had to choose between continuing to lend to the state-owned companies or watching those state-owned companies' loans become worthless as the companies collapsed for lack of funding. The latter choice was not only unattractive, since most bank assets were in the form of notes payable from the state enterprises, but also implausible, since those bank clients were also generally major shareholders in the banks.[501] Continuing to lend to the existing clientele was not only difficult to avoid, but it was easy and familiar business for the bankers involved.[502]

As a result, there was very little credit available for entrepreneurs, including those who wanted to borrow money to participate in privatization of other companies. The policy also hindered privatization by inhibiting the foreign investment of firms that saw a hopelessly outdated banking system as a problem for them.[503]

Privatization of the Banking Sector

Privatizing the banks would certainly benefit the economy, but that was not the immediate impetus for the privatization plans that began to be formulated in 1990. Rather, the banks were in trouble and needed infusions of capital. Estimates of the level of bad debts in 1992 ranged from $1.2 billion to $3 billion.[504] The State Banking Supervision estimated in October 1992 that 21 percent of all loans held by Hungarian banks, representing 12 percent of all bank assets, could be classified as "high risk."[505] Both poor management and the problems inherited from the HNB when the commercial banks were spun off were causes of the banking crisis. But the most fundamental cause was more general: the aggregate economy of Hungary was in terrible shape and getting worse. The State Banking Supervisor suggested in late 1992 that the banking crisis was directly related to the general economic conditions in the country.[506] That was unquestionably true. And yet the banks were different from the economy as a whole in that the commercial banks were financially tied to the worst sector of the Hungarian economy: the state sector.[507]

Regardless of causes, the problems at the banks were acknowledged both inside[508] and outside government[509] as the primary motivation for privatizing the banks. The financial difficulties did not lead to insolvencies until the summer of 1992, but the problems were evident several years before that. Even if some of the banks did not face the imminent possibility of collapse, it was widely recognized that those banks would need substantial additional capital if they were to be capable of assisting in the finance of economic reform.[510]

The economic reality necessitating the privatization of the banks resulted in several laws mandating that privatization. A new banking act was passed in 1991 forbidding any entity other than another banking institution from owning more than 25 percent of any Hungarian bank after January 1, 1997.[511] It also mandated the reduction of voting rights for the state to no more than 25 percent by January 1, 1995.[512] In addition, the same law required as a practical matter an influx of capital at most banks by requiring them to achieve and maintain certain minimum capital levels and set aside specified risk reserves.[513] The Hungarian National Bank was required to divest itself entirely of bank shares within two years pursuant to a different law.[514]

Still, privatization was a delicate political issue at the time. Some officials advocated allowing foreigners to control small banks and take minority stakes in large ones; others objected to privatization to foreigners at all.[515] The government made assurances in 1991 that the largest savings bank (OTP) would not be privatized and that the four largest commercial banks would remain Hungarian-owned even if they were privatized.[516] Eventually, this policy was modified to provide that the state could retain a majority of OTP shares, and permission of the HNB would be required before any foreign entity purchased more than a 10 percent stake in any existing Hungarian bank.[517]

Although the process was begun in 1990 as soon as it became clear that the banks would not have sufficient capital to support economic reforms,[518] there was very little activity toward privatizing the banks until 1991. That year laws were passed permitting foreigners to set up foreign-owned banks and to purchase interests in existing Hungarian banks.[519] In February 1992, HNB president Peter A. Bod told reporters that he hoped the process would begin in 1992, although he confessed that it would be at least 1993 before the process would be completed.[520] That April, the government put forward a five-point plan for privatizing the banks:

1. Privatization would come about through capital increases, not just purchase of outstanding shares.
2. The state, not the banks, would make the decisions concerning when and how to privatize.
3. The state would remain a major shareholder, down to the 25 percent level mandated by law as of January 1, 1997.
4. To the extent possible, banks would remain majority owned by Hungarians, at least with the big banks.
5. The state would try to give small shareholders an opportunity to buy shares.

Approximately one week later, five-point plan's author, Minister Without Portfolio for Privatization Tamas Szabo, established a committee to oversee government policy on privatizing the banks. Included on the committee were representatives of various ministries and the Hungarian National Bank.[521] The process was aided by an SPA directive called the SPA Fund Transfer Policy, which required a state-owned enterprise to turn over to the SPA any bank shares held by it when it sought registration under the 1988 Act on Associations.[522] The immediate effect would be a concentration of shares in the hands of the state, but this ironically would facilitate privatization, since outstanding shares could be purchased directly from the state.[523]

The SPA also showed that it was willing to take tough measures to ensure its control of the bank privatizations. On April 30, 1992, the SPA went to a meeting of the board of directors of the Hungarian Credit Bank, the second largest bank in Hungary. Holding the state's shares, the SPA represented a majority of those present. It used that position to sack the entire board of directors and the supervisory committee, replacing each person with an SPA nominee. SPA managing director Lajos Csepi explained only that the action would further the privatization of the bank.[524]

Still, progress toward privatization was slow in 1992. There were a lot of inquiries from potential bidders.[525] For example, the European Bank for Reconstruction and Development announced in September 1992 that it was ready to invest in Hungarian commercial banks and was negotiating with the government to do so.[526] But there was little real action. One problem was that too many organizations in Hungary wanted a say in what course privatization would take.[527] The Finance Ministry, the Hungarian National Bank, the SPA, the State Banking Supervisor, the banks themselves, and several parliamentary committees all had plausible claims to jurisdiction.

There was also evidence of bureaucratic confusion in the process. In early September 1992, Deputy State Secretary Imre Csuhaj announced that the state would choose a single adviser for the privatization of all banks.[528] This caused an uproar among the bankers, who had thought they would each choose their own strategy and adviser.[529] Almost immediately, SPA Deputy Director Karoly Farkas said, rather diplomatically, of Csuhaj's statement, "there must have been some misunderstanding." He explained that one adviser would draw up a scheme for credit consolidation mechanisms to improve the financial positions of all the banks before privatization. But each bank would then be entitled to choose its own adviser from a list approved by the government.[530]

Later in September 1992, the government published a short list of eight possible advisers to compete for the four positions as advisers for Hungary's four largest banks. In addition, one of those four would be chosen as government adviser on general issues of privatization of commercial banks.[531]

Although none of the large banks was privatized by the end of 1992, several smaller banks were. For example, France's Banque Indosuez bought 63.8 percent of the Hungarian Kulturbank, and Luxembourg's Fransaholding purchased another 11.2 percent of the same bank.[532] In addition, Wiener Allianz Versicherungs-Aktiengesellschaft acquired 16 percent of Hungary's Posta Bank, which had the greatest market dispersal with branches in every post office in Hungary.[533]

But in at least one respect, the Hungarian government was moving in precisely the wrong direction: in the *de*privatization of Konzumbank. Konzumbank was small and privately owned. It served as one of the few sources of private capital for small start-up businesses. Unfortunately, the bank was badly managed and engaged in unsound lending practices in the late 1980s and early 1990s. By late 1992, Konzumbank was in danger of collapsing. Rather than simply let the market discipline this imprudent economic actor, however, the SPA took 75 percent of Konzumbank, while expressing the hope that it could be reprivatized later.[534] Thus, at the same time it was attempting to privatize most of the banking sector, the Hungarian government was nationalizing part of it. Clearly, the bulk of bank privatization would come after 1992.

The Ybl Bank Collapse and Other Scandals of 1992

Although its effects are difficult to estimate, one of the factors that may have slowed bank privatization in 1992 was the publicity surrounding certain bank scandals and the collapse of three banks during the summer of 1992. The most highly publicized bank insolvency was that of Ybl Bank.

There had been rumors all summer that some of the banks were in dire straits, and many of the banks stocks were unusually active.[535] Still, shock waves reverberated through the financial community of Budapest when on June 23 the State Banking Supervisor announced that it was forbidding Ybl Bank from accepting deposits or engaging in active financial operations.[536] Six days later the bank itself announced it was voluntarily suspending its securities trading activities under the cloud of investigations by both the State Banking Supervisor and the State Securities Supervision Board.[537] On July 3, the sssb announced that its investigation showed trading violations by Ybl Bank, and on July 7, the sbs formally suspended the bank's trading privileges for numerous alleged violations of the banking regulations.[538] At the same time the sbs put the bank into receivership by putting the bank under the control of a board of supervisory commissioners.[539] And less than two weeks later, the Budapest Stock Exchange formally suspended the bank's membership there.[540] By the end of July, Ybl Bank was formally declared insolvent by the State Banking Supervisor, and its directors were charged with fraud.[541] The firm of Deloitte and Touche was appointed liquidator.[542] As of the end of 1992, no buyer for the bankrupt bank had been found.[543]

What happened at Ybl Bank? In 1993, the Budapest chief presecutor formally charged the chairman of the board, the director general, and the deputy director general of corruption and banking malpractice leading to the loss of 6 billion HuF (approximately $80 million).[544] According to the chief prosecutor's office, the alleged offenses stemmed from the multiple discounting of bills of exchange and bribery related to bills of exchange.[545] One major source of losses alleged was the granting of unsecured loans to the major investors in the banks.[546] With the bank insolvent and no deposit insurance, depositors pursued legal claims against the State Banking Supervisor, alleging failure to supervise adequately.[547]

Whatever the nature, corruption at the major banks was allegedly rampant in the early 1990s. As the head of the State Banking Supervisor put it, "You don't need a gun to rob a bank in Hungary.... It can be done more simply with financial tricks."[548] At approximately the same time the Ybl Bank problems were coming to light, two other banks were being put into receivership as well. Like Ybl, the General Bank of Venture Financing (AVB) and the Gyomaendrod Savings Bank (GSB) came under State Banking Supervisor scrutiny in the summer of 1992.[549] In both cases, fraud on the part of the managers was blamed for the collapse, and in both cases the State Banking Supervisor appointed supervisory commissioners.[550] GSB was ultimately liquidated, but a majority of AVB was sold to a German bank.[551] As was the case with Ybl, investors in AVB and GSB allegedly took out large unsecured loans and failed to repay them.[552] The State Banking Supervisor charged the directors of GSB with fraud.[553]

Following an investigation that continued into 1993, police discovered that the pattern of fraud alleged against managers at Ybl, AVB, and GSB was repeated throughout Hungary's banking sector, and prosecutions were numerous.[554] In all, fraudulent practices were alleged at 25 to 30 banks in various parts of Hungary, with the highest concentrations in the rural areas.[555] The Hungarian National Bank reported to Parliament that most of the cases involved the taking of bribes for the granting of loans.[556] Estimates of depositor losses stemming from the fraud ranged from 16 billion HuF ($213 million)[557] to 20 billion HuF ($267 million).[558]

The scandal may have set back attempts to privatize the banks by putting the industry as a whole into disrepute. Ironically, however, in the case of AVB, the alleged fraud may have facilitated that bank's privatization. The example of AVB makes a nice case history of the problems in both the Hungarian banking industry and the privatization of that industry.

AVB was initially established by a group of approximately two dozen state-owned enterprises that wanted a source of venture capital. The bank was unprofitable almost immediately, however, as a result of unsound lending policies. It charged extremely high interest for its loans, but the projects were so risky that the interest premium was insufficient to make up for the risks.[559]

Despite these problems, Westdeutschelandesbank (WDLB) showed an interest in AVB as early as 1991 as a result of its desire to establish a foothold in the Hungarian banking industry. At that time, German firms were investing heavily in Hungary, and it would be natural for them to want to do business with an affiliate of a German bank. In addition, WDLB wanted to establish a presence among the growing Hungarian entrepreneurial sector.[560]

With the approval of AVB, WDLB began acquiring shares of AVB in the over-the-counter market. In early 1992, the State Property Agency approached WDLB with the proposal that WDLB acquire a majority interest in AVB. Later that year, WDLB did in fact acquire 58 percent of AVB's stock, half directly from the SPA and half from those state-owned enterprises still owning AVB shares. Nineteen percent was owned by a consortium of small companies known collectively as the Autoklinik Group. Of the remaining shares, 7 percent were held by the state's social security fund, and the rest were spread among smaller institutional and individual shareholders.[561]

After acquiring the majority of AVB's shares, WDLB discovered problems in the bank. Most importantly, according to a WDLB senior officer, WDLB discovered irregularities in the loans to those companies forming the Autoklinik Group. The loans had not been examined closely by WDLB before it purchased the AVB shares, since each fell under the size threshold WDLB set for examination of individual loans. WDLB did not discover until after its acquisition that those loans were really a package, lent to a consortium of companies that owned a substantial minority stake in the bank. They turned out to be uncollectable. Taken together and combined with the other uncollectable loans held by AVB, they were enough to render AVB insolvent just after WDLB's majority acquisition. In the summer of 1992, the banking supervisor stepped in and forced AVB to cease doing business.[562]

AVB went into bankruptcy proceedings and restructured itself with the help of the SPA and with a 30 percent reduction in its debts by its creditors. But the new managers of AVB discovered the need to restructure much more than AVB's financial structures. The books had been kept by hand according to principles not recognizable to Westerners. In addition, WDLB's hopes of gaining the business of Hungary's entrepreneurial class were dashed when those hopes came into conflict with conservative German lending practices: small Hungarian business were seen to be much too risky. As of the spring of 1993, bank officials projected that 60 to 70 percent of future business loans would be to German companies operating in Hungary and 20 to 30 percent would be to other Western European businesses. Lending to Hungarian firms would be limited and would no longer take the form of venture capital without substantial collateral.[563]

Thus the privatization of AVB, coming as it did just as the banking scandal was coming to light, represented a mixed blessing from the Hungarian perspective. On the one hand, without WDLB's acquisition, AVB probably would have been liquidated, meaning that its creditors would never have been paid and its 144 employees would have been out of work. On the other hand, the acquisition by a German bank meant the elimination of a source of badly needed venture capital for Hungarian entrepreneurs. AVB's capital would be used to promote German and other Western European businesses, not Hungarian businesses. From the acquirer's perpsective, the AVB story is a cautionary tale about the hidden difficulties of making acquisitions in so unsettled a system.

The 1992 Bank Bailout

In 1992, with such a high proportion of its loan portfolio in the high risk category, the private Hungarian banking system was in danger of collapse. The destruction of commercial credit facilities would have been bad enough. But in Hungary at the time, there was no deposit insurance. Widespread bank closures would have meant the loss of savings for much of the population, which could in turn have wiped out public confidence in and support for private economic institutions and a private economy generally. For those reasons, the government, acting through the Hungarian National Bank, moved to shore up the banking system in late 1992.

The scheme invented was designed primarily to remove from the banks the pressures placed on them by their inheritance of bad loans from the Hungarian National Bank when the latter spun off its commercial banking functions and the aggregate economic decline in Hungary over which the banks obviously had no control. The plan, which was approved by the government in mid-December 1992, worked like this:

A new government agency, the Hungarian Investment and Development Company (MBF Rt.) was established. Banks could trade in their nonperforming loans to MBF, and MBF would give them in return new government bonds with a 20-year term and floating interest rate. Thus bad assets were replaced with good ones. But only loans made before October 1, 1992, could be exchanged. Loans from 1992 could be exchanged for bonds worth 80 percent of the face value of the loans; pre-1992 loans could be exchanged for bonds worth 50 percent of the face value of the loans. Any exchanges had to be made before the end of 1992. In addition, any banks taking part in the exchange program had to agree to pay part of their profits into a fund to help finance the program.[564]

Privatizing the Agricultural Sector

At least in the banking sector, policymakers' problems were limited to legal and economic difficulties. Privatization of the vast farm holdings of the state was complicated by both politics and the special position of agriculture in Hungary's economy. Agricultural land occupies an important emotional/nationalist position in Hungary as in many countries. Nineteen percent of Hungary's work force is employed in farming.[565] Privatization through mass sales to foreigners could have been politically untenable and could have jeopardized the already ambivalent public support for market reforms. In addition, one of the member parties of the ruling coalition, the Independent Smallholders party, focused on a nationalist, populist, and agrarian agenda and probably would have bolted from the coalition if the MDF began selling farmland to foreigners on a massive scale.

Privatizing the agricultural sector was also different because this was a sector that had been successful for an extended period. Hungary had long been a net exporter of agricultural commodities, many of the exports going to hard-currency countries. Heavy subsidies kept production high and prices low, resulting in full shops and low food prices. As the government began dismantling subsidies in the first years of the 1990s, it was clear that food prices would have to rise, and with the rise in prices would come intense public scrutiny of government policies. Hungary could ill afford making a mess of privatizing this sector.

As early as 1990, it was clear that there would be problems in privatizing the farming sector. Hungarian agricultural policy maintained artificially low prices by subsidizing overproduction rather than emphasizing efficiency in operations.[566] Privatization and removal of subsidies would cause prices to rise and supplies to diminish. And the ownership confusion that plagued other privatizations was particularly acute in the farming sector.[567] In 1992 there were 122 large state-owned farms[568] representing approximately 10 percent of the arable land and 10 percent of production.[569] The other 90 percent was made up of a small quantity of privately owned land and vast tracts of land owned by cooperatives. The cooperatives were legal entities whose owners nominally were the various entities with interests in the production of the farms, mostly suppliers (e.g., machinery producers) and customers (e.g., food processors and trading companies). These suppliers and customers were themselves often owned either directly or indirectly by the state, with confusing ownership patterns. The state ultimately controlled the means of production, but the concept of ownership in the sense of legal title was not particularly important and had become terribly confused over time. In addition, until the compensation issue was settled in 1992, no one could be certain of the rights of the former owners and their heirs from whom the land had been appropriated in the 1940s and 1950s.

Not surprisingly, the first privatization efforts focused on those 122 state farms, which were clearly owned by the state. Even there, however, the problems were numerous. The Monor farm, just outside of Budapest, offers an example. This 4,400 acre farm was in 1992 in the process of being privatized with the assistance of the United States Agency for International Development (USAID).[570] In the late 1980s, it had been a prosperous farm with profits ranging from $666,000 to $1.3 million annually; it employed approximately 1,000 workers.[571] Sales were largely to the Soviet Union, and prices were subsidized. By 1992, farm production in Hungary was down approximately 20 percent. The Soviet Union was gone, and subsidies had been slashed. Western Europe, with its own protectionist policies, would not buy the Monor farm's surplus, and the remaining customers for Monor's production (mostly domestic food processors) were not paying their bills. Approximately 300 workers had been cut by the spring of 1992.[572]

The Monor farm was a prime candidate for privatization. But when the SPA set about privatizing the farm, problems were sprouting with more vigor than the corn. The problems of indistinct ownership and possible claims by former owners were chasing away many buyers. In addition, many of the workers had allocated to them small plots of land on which they could grow produce privately. These plots were not "owned" by the workers, but had been tilled by them long enough for the workers to have become dependent upon them and to assert ownershiplike interests in the plots. Finally, the Monor farm, like most in Hungary, was overly vertically integrated. It raised corn, produced its own seed corn, milled corn, and baked corn muffins for retail sale all at the same facility. The problem was that most modern agricultural enterprises concentrated on only one sphere and did not want to buy such a vertically integrated enterprise.[573] At the end of 1992, the Monor farm was still struggling and was still unsold.

The political and practical difficulties of privatizing the farming sector were reflected in an ambivalent government policy. The government announced in May 1991 that some state farms would be held permanently by the state instead of being privatized.[574] In addition, its draft privatization policy, floated in June 1991, made clear its intent to sell most arable land to individual Hungarians.[575] Finally, although draft legislation being considered in 1992 would permit almost unrestricted real estate purchases by foreigners, that legislation would make ownership of farmland impossible for foreigners. That legislation was still pending late in 1992.[576]

Seven

Another Mid-Course Correction—1992 and the Shift Toward the Domestic Market

Courting the Domestic Investor

One of the things that became clear in 1991 and early 1992 was that the SPA was first and foremost pragmatic. Although there were certain principles that guided and constrained SPA policy, and although political realities channeled that policy, the SPA demonstrated a willingness to remain flexible and to try different programs to meet changing circumstances. In mid and late 1992, the SPA continued trying different techniques in an attempt to speed up privatization, but a change in course was evident. Whereas in 1991 and early 1992 the changes in programs represented trial and error and attempts to make different types of state assets saleable to foreigners, in mid and late 1992, changes began emphasizing domestic participation.

There were two primary reasons for this shift. The first, and the one heard publicly the most, was a desire to contend with the political difficulty of selling Hungarian assets to foreigners.[577] As the head of the SPA said in early 1992, "For social and political reasons, it is important to raise the proportion of domestic participation."[578] But the other reason, largely unspoken, was probably the true motivation: the SPA was running out of foreign buyers and needed new blocks of customers to keep the privatization efforts from falling even further behind schedule.[579] In October 1992, Hungarian economist Gyorgy Szakolszay said at a joint conference of Germany's

Konrad Adenauer Foundation and Hungary's Privatization Research Institute that it would take 5,000 years to privatize all state property using present techniques at the present pace. This may have been hyperbole, but the goal of 50 percent by 1994 was, by the end of 1992, becoming increasingly unrealistic. In the two years ending in September 1992, one estimate suggested that only approximately 10 percent of state assets had been privatized.[580]

Whatever the reason, in mid-1992 the Hungarian government took a number of steps to facilitate the acquisition of state assets by domestic Hungarians.

The Voucher Taboo

This shift toward the domestic market was made a little difficult by the rhetorical corner into which the government had painted itself by emphasizing its refusal simply to give state assets away. Many observers both inside and outside Hungary advocated at various times the gratis distribution of state assets to the Hungarian populace as a primary privatization technique.[581] The voucher system of Poland, whereby individuals were given scrip that could be traded for state assets, was often held up as a model for Hungary.[582]

Justifications for such giveaways were numerous. Some commentators focused on the need to create quickly a private economy and concomitant middle class. Others argued that once foreigners had snapped up the most attractive assets, there would still be far too many state assets for domestic savings to absorb.[583] Domestic demand was indeed low, and there was some evidence to support this latter argument that Hungarians simply could not afford to buy shares in sufficient quantities to be significant macroeconomically. A blue ribbon commission on the Hungarian economy estimated in 1990 that the total domestic savings of Hungarians was only $300 million.[584] By one estimate, domestic savings could purchase only approximately 10 percent of state-owned assets even if credit were available.[585] Other statistics released by the Hungarian National Bank show that by August 1992, Hungarians had net domestic savings of approximately $10 billion.[586] And according to an estimate by SPA official George Hollo, Hungarians in 1992 held more than $1.8 billion in hard currency savings abroad.[587] This could be used to buy shares if Hungarians were interested in buying shares. But regardless of whether Hungarians had sufficient savings to participate in privatization, they did not do so in significant numbers. And they did not primarily for two reasons unconnected with their aggregate savings. Second, even if Hungarians had the funds to invest, they did not have the ability to pool their money into investment funds until 1992, when a law authorizing such funds was passed. Since almost all the privatizations were private placements and the savings of Hungarians was widely dispersed, aggregate savings figures did not mean Hungarians were able to participate significantly in the process. Third, and perhaps most importantly, these data do not tell us anything of the willingness of Hungarians to invest in privatized firms. The Hungarian middle and professional classes tend to save their excess income in the form of hard assets. That is, although many have the ability to buy shares, they lack interest. When asked where they invest their money, more than one Hungarian told me he or she improved his or her housing, perhaps by adding a new room or upgrading the quality in an existing house or by trading up to a more expensive flat. This preference for investing in useful hard assets reflects the pervasive Hungarian uneasiness about an unpredictable future and the almost universal unfamiliarity with the concept of stock ownership. Also, with interest rates on bank deposits in excess of 25 percent, Hungarians have little incentive to buy shares.[588] When asked what it would take to interest Hungarians in share purchases, one financial consultant put it this way:

> They would have to be sure of making a lot of money very fast, because they do not believe in the future. They do not know if this company will exist in two years. They do not know what the government will do. Also, they can make more money simply by putting money in the bank.[589]

All these facts have resulted in a disappointing level of interest on the part of Hungarians in share purchases.

Yet despite this lack of interest and the calls from some quarters to emulate the Polish voucher system, the Hungarian government resisted publicly, clearly, and often. Indeed, in 1989, when privatization efforts were just beginning under the HSWP, Dr. Janos Martonyi, the commissioner in charge of privatization at the time, made it clear that the government was going to be in the business of privatizing, not reprivatizing.[590] In other words, Hungary would sell state assets, not redistribute them for free.[591] This point was made repeatedly by the SPA throughout the early 1990s.[592] The giveaway voucher system of Poland was a special subject of derision for Hungarian government officials.[593] Use of the term *voucher* was considered taboo.[594]

That rhetoric somewhat limited the government's options when it heard grumblings about foreign domination of the privatization process and when the pace of privatization did not meet expectations because of a lack of purchasers. How does a government, having sworn to sell (not give away) assets, privatize an economy and ensure domestic ownership if there are not enough buyers willing to pay acceptable prices and domestic investors are sitting on the sidelines? This was the question facing policymakers at the SPA in the early 1990s.

But These Aren't Vouchers

The first response was to come up with a wide variety of schemes that the government hoped would boost domestic demand. The government had been trying many of them from the first days of the SPA's existence, but the attempts to increase Hungarian participation in the process clearly accelerated as time went by.

In the early months of 1990, just after the SPA came into existence, attempts were limited to education programs. The government ran educational advertisements in the local media explaining privatization, the basics of company structures, share ownership, and the like, and encouraging citizens to buy shares in companies being privatized.[595] The Austrian investment firm Girozentrale even sponsored a newspaper game in which the reader/player was engaged in portfolio management.[596]

The early attempts had little effect on domestic demand. As a result, the government began offering other incentives. Their variety is further evidence of the pragmatic and creative quality of the MDF's privatization policy. The various plans can be categorized into three large groups: preferential loans for domestic participants in privatizations, special preferences and incentives for employees and managers of privatized firms, and special favorable consideration for domestic bidders in the competitive bidding processes.

Preferential Loans for Domestic Purchasers. The Hungarian government experimented with numerous mechanisms for easing credit for those domestic investors who were interested in buying privatization shares. In September 1990, the SPA announced that the government was in the process of establishing two funds to lend to the small investor.[597] The first was the Existence Credit (E-Credit). The original idea called for the E-Credit fund to be set up with both Hungarian and German capital, and its monies were to be used to provide low-interest loans to Hungarians wishing to purchase privatization shares.[598] The initial capitalization was to be 184 million DM, with 100 million DM in the form of a loan from the German government and the other 84 million DM in the form of a loan from the Hungarian National Bank (HNB).[599] But by the time the fund was set up at the beginning of 1991, the Germans had backed out. They saw (almost certainly correctly) that the future of the Hungarian economy was in entrepreneurial start-ups rather than privatization. Thus, the Germans wanted the E-Credit to be available for such start-ups. As a result of the rift, the HNB set up the fund on its own, establishing preferential credits through the commercial banks.[600]

Under the terms of the E-Credit, a sliding scale for terms was established whereby the more one borrowed, the more favorable the terms became. Investors could borrow up to 50 million HuF ($667,000), with the down payment ranging from 15 to 20 percent and the amortization on a schedule from 6 to 10 years. The interest rate was set at 75 percent of the HNB's base rate, although banks were allowed to charge as much as 4 percent as a commission.[601]

The HNB also established in 1991 the Privatization Credit, which was set up to help small business owners purchase state assets.[602] Both funds were, by most measures, flops.[603] The Privatization Credit made only four loans in 1991, totaling a little over $150,000. The E-Credit loaned out only about $20 million.[604] These disappointing results occurred almost certainly because domestic investors were not really interested in buying shares even on preferential terms and the interest rates were still too high — approximately 20.5 percent on the E-Credit in 1991.

In early 1992, the government moved to improve the two credits in four ways. First, it merged the two funds.[605] Second, it eliminated the upper loan limit of 50 million HuF.[606] Third, it lowered the interest rate from 75 percent of the HNB's base rate to 60 percent of that rate.[607] That pulled the interest on the E-Credit loan down from 16.5 percent to 13.2 percent by April 1992.[608] Then in December 1992, the government announced that it was dropping the rate to between 6 and 7 percent and extending the amortization schedule from a maximum of 10 years to a maximum of 15 years.[609] The new terms would, significantly, apply to outstanding loans as well as new ones.[610] Finally, the government expanded the circle of the those eligible for the credits by expanding the uses to which they could be put.[611] Henceforth, E-Credits would be used not only to buy privatized shares, but also to start new businesses[612] and even to buy new flats.[613]

With the disappointing results of the E-Credits having become apparent, the government moved in late 1992 to broaden the availability of preferential privatization loans in a couple respects. First, in early October, SPA chief Lajos Csepi announced that the government planned to distribute credit cards to Hungarian citizens that would carry the same rate of interest as that paid by the government to the HNB for credits (between 12 and 14 percent in 1991 and 1992). These credit cards would then be used to buy shares.[614]

Two days later the government announced that in the near future all Hungarians would get "loan coupons" up to a maximum value of 1-2 million HuF (approximately $13,300 to $26,600), which could be used to buy privatized shares. The loan coupons would require only a 1 or 2 percent down payment and would carry interest rates at 40 to 50 percent of the HNB's base rate, with terms to ten years. The government expected privatization revenues to be boosted approximately 70 billion HuF (approximately $93 million) in 1993 as a result.[615]

Special Preferences for Workers and Managers of Privatized Firms. Employment dislocations resulting from efficiencies brought about by privatization were a constant concern for government policymakers seeking to ensure the continued political viability of privatization efforts in Hungary in the early 1990s. That fact, combined with the desire to increase domestic participation in privatization generally, resulted in special policies that enabled workers and managers to benefit directly from privatization by allowing them to acquire shares under preferential terms.

In 1991 the government promulgated Policy Guidelines on Assets to guide the SPA in its decision making. Among its many provisions were terms that permitted employees and managers who were submitting bids for privatized assets to apply for special preferential treatment before the SPA.[616] The SPA was, in fact, authorized to grant unlimited price or installment payment concessions to employees.[617] The example of the Kanizsa Brewery illustrates how this advantage worked in practice.

In 1992 the Kanizsa Brewery held 12.5 percent of the Hungarian beer market, employing 1,070 persons. Following its transformation to a share company, 100 percent of the shares were owned by the SPA. When the SPA sought to sell its holdings,[618] it received three bids: one from a foreign consortium, one from a group of Hungarian investors, and one from a group representing 420 of the Kanizsa employees and managers. Against the advice of its adviser, Morgan Grenfell, the SPA awarded Kanizsa to the employee group even though that group's bid was not the highest. Indeed, the group's bid called for a highly leveraged transaction, with only approximately $400,000 in cash.[619]

Later that year, in November 1992, the SPA announced that it would make available the entire state holding in AGROINVEST Rt. for purchase by that company's employees.[620] At the time, the state owned 95.9 percent of AGRONINVEST's shares, with the rest divided among the employees and local councils. The SPA announced that it would sell the state's shares to any interested employees at 80 percent of the face value of those shares.[621]

The government was also taking legislative steps to bolster employee purchases of shares. During the summer of 1992, the Hungarian Parliament passed Act XLIV of 1992 on the Employees' Part-Ownership Program, authorizing and regulating the establishment of Employee Share Ownership Plans (ESOPs). In the preamble to the act, the Parliament was quite clear concerning its motivations: "for accelerating privatization and in order to promote the ability of employees to acquire . . . ownership shares in the economic association employing them."[622]

Under the act, any employee of a Hungarian company who works at least half-time for that company and has been employed there at least six months (or up to five years if the program managers decide to increase the threshold) is entitled to participate in the Employee Part-Ownership Program.[623] The program comes into existence if at least 25 percent of the employees express a desire to acquire shares through such a program.[624] Once set up, the program is entitled to purchase, with E-Credits and other preferential loans, shares in the company.[625] Periodic payments for the shares (and concomitant payments by the employees) are to be made pursuant to the bylaws of the program.[626] Income on the shares held by the program is tax exempt, and the company receives a tax deduction for the value of shares transfered to the program.[627] The act was passed by the Parliament on June 9, 1992, and received a lot of attention in the Hungarian press.[628] Its effect on privatization and domestic participation could only be judged after a couple years of operation had passed.

Preference of Domestic Bidders Over Foreign Bidders. At about the same time the SPA was reviewing the bids for the Kanizsa Brewery and implementing its policy favoring employees and managers in the competitive bidding process, in May 1992 the government announced a new policy. In cases of similar bids, the SPA would prefer domestic investors over foreigners.[629] The announcement followed publication of opinion polls showing that fully 34 percent of Hungarians opposed selling state industries at all (as opposed to 39 percent in favor), partly no doubt because of the domination by foreigners of the process.[630] Nine of the ten largest privatizations in 1991 had gone to Westerners.[631]

The next month a bill was introduced in the Hungarian Parliament that would require the SPA to prefer domestic bidders.[632] There were of course precedents for such preferences in practice. In the privatization program whereby small shops and restaurants were auctioned off, only domestic bidders were permitted to participate. And, when the liqueur manufacturer Zwack Unicum was privatized, the bid submitted by Peter Zwack, a descendant of the original manufacturing family, was preferred over that of liquor giant Seagrams, largely on emotional grounds.[633] But formalizing the preference was a new development that further evidenced a shift away from the MDF coalition's purported exclusive emphasis on gaining the highest possible price for privatized assets.

Indeed, the two most prominent privatizations of 1992 both involved reservation of a percentage of shares for sales to domestic investors on preferential terms.[634] The first involved Pick Szeged, the world-famous manufacturer of fine Hungarian salami. The transaction was both sizable and highly visible due to the perception of the firm's quality as an investment. Pick Szeged had earned in 1991 approximately $9 million on sales of approximately $138.6 million.[635] While the deal was still being formulated in July 1992, the SPA told reporters that it was considering several possible incentives for domestic purchasers of shares, including price discounts of up to 10 percent, deferred payment plans, and even bonus shares that investors would receive for free.[636] In addition, 5 percent would be reserved for the local councils, 10 percent for the employees, and another 20 percent for compensation coupons.[637] In all, the SPA was aiming for approximately 70 percent of the shares to go to domestic Hungarians.[638] When public sales began on the morning of October 27, the allocation to various buyers illustrated the degree to which the SPA was favoring domestic purchasers:

    5% + 1 share — reserved by SPA
    10.8% — employees on preferential basis
    4.1% — local councils
    20% — compensation coupons
    11% — banks, at discount, in debt/equity swap

The rest would be sold to the public, but Hungarian individuals were entitled to purchase at a 10 percent discount so long as they held the shares for at least one year.[639]

A similar pattern emerged when the Danubius hotel chain was finally privatized in late 1992. As discussed above, the sale's on-again/off-again pattern frustrated those involved.[640] Finally the issues that had held up the privatization were resolved as best they could be (in part by the exclusion of the Gellert Hotel,[641] whose ownership structure was a particular mess,[642] from the privatization), and the sale commenced. As with the Pick Szeged transaction, small domestic investors were sold shares on highly preferential terms.[643]

In addition to these attempts to bolster domestic participation, other ideas were floated as well. SPA chief Lajos Csepi, for example, told reporters in January 1992 that he would like to see tax breaks for privatization share purchases.[644] In addition, the SPA continued to make some shares available for compensation coupons and in May 1992 announced that it was setting up an investment fund that would hold interests in various privatized firms (seven to begin with). The shares of the investment fund would then be available for purchase with compensation coupons.[645] That action was probably motivated both by a desire to increase domestic ownership and by a desire of the SPA to dump some of its residual holdings in privatized firms or firms that were not attracting potential purchasers.

In December 1992 the government's cabinet officially approved the shift in emphasis away from sales to foreigners for cash and toward the domestic market. That shift toward the domestic market would mean further inducements such as easier credit.[646]

Although some of these attempts by the government to increase domestic participation were cost-free to the government, others (e.g., low-interest loans, sales on preferential terms) clearly acted as subsidies. Thus the government was obviously weakening in its resolve to sell—not give away—state property: there is no economic difference between a subsidized sale and a partial giveaway. The distinction the government attempted to maintain between giveaways and subsidized sales began to crumble in October 1992, when SPA official Tamas Szabo announced that a voucher system based on the Czech model would be implemented sometime in 1993.[647] This came on the heels of publication of figures showing that 85 to 90 percent of all privatization revenues in 1991 came from foreigners,[648] a situation that caused the SPA managers to instruct the staff in August 1992 to be more aggressive in cultivating domestic involvement.[649]

Although this announcement of a voucher system seemed a clear reversal of government policy, Szabo made clear that the instruments used to give state property away would not be called "vouchers" but would look to observers like those used in the Czech system.[650] And SPA officials were privately continuing to insist that there was no change in policy: these were not going to be vouchers.[651] It was a telling moment in Hungary's short privatization history. The move first toward greater domestic incentives, then to more direct giveaways demonstrated certain facts that after 1992 would continue to color Hungary's privatization efforts:

1. Foreign interest in purchasing privatized assets was, after an early euphoric period, wearing thin. This was partly a result of the discovery by Westerners that Hungary was not the gold mine it was thought to be. Rather, they discovered that reaping profits from Hungary would require patience and a long-term investment strategy. The decrease in foreign interest was also caused in part by the fact that the best Hungarian assets were identified and purchased early on, leaving less attractive assets. Third, foreigners became increasingly frustrated with the SPA's cautious approach and the difficulty of sorting out ownership interests in the various assets. The interest of foreigners rose from 1989 through 1992,[652] but it seemed to be on the back side of the curve by the end of 1992.

2. Domestic Hungarians were still inhibited from participating in privatization sales, both because of their inability to participate and because of their unwillingness to take the risks that equity participation demands. To the extent that Hungarians were willing to

forgo the high interest rates paid on bank deposits in favor of a more risky venture, most were more likely to start their own businesses.

3. The MDF coalition was conscious of the need to maintain popular support for its programs.[653] At the same time, it needed to ensure the continued support of its own populist/nationalist wing, which was increasingly vocal as the 1990s progressed. Both these needs were satisfied in part by ensuring greater domestic participation in the privatizations and countering claims by some that Hungary was being sold cheaply to rapacious foreigners.

The Status of Privatization Efforts at the End of 1992

All in all, 1992 turned out to be a good year for the SPA and for privatization efforts generally. Many of the devices to accelerate the process had underperformed in comparison with expectations, but a lot of state property was sold in 1992. In fact, in 1992, twice as many state companies were privatized as in 1990 and 1991 put together.[654] Privatization brought in approximately $780 million in 1992,[655] about 70 percent of it in the form of much-needed hard currency;[656] this was much more than the $546 million projected in January 1992.[657] By the end of the year, the private economy accounted for nearly one-third of Hungary's GNP.[658]

Most of the credit for that figure, however, goes to the expansion of the entrepreneurial sector and the collapse of the state sector rather than to privatization. In the two and one-half years ending in December 1992, only 2.5 to 3 percent of the 2,000- plus state-owned companies had been fully privatized.[659] At the same time, the percentage of the economy in private hands grew from less than 10 percent to more than 30 percent. In 1991 alone, 431,000 new one-person businesses were formed in Hungary and 136,000 new companies were registered.[660] There can be no question that the growth of the private economy was "privatizing" the overall economy more than were SPA privatization efforts.[661] Thus, other macroeconomic events were more significant than privatization. In 1990 the 2,000-plus state-owned enterprises were valued variously at figures ranging from approximately $26.7 billion to approximately $37 billion.[662] But privatization revenues for the three years combined totaled only

1992 and the Shift Toward the Domestic Market 107

approximately $1.2 billion,[663] or no more than 4.5 percent (and perhaps as low as 3.3 percent) of all state assets if the 1990 calculations of value are used.

But they cannot be used, because they were based on book value. At the end of 1992, the SPA was estimating that the approximately $1.2 billion worth of assets privatized to that point were approximately 20 percent of all state-owned business assets.[664] SPA chief Lajos Csepi made clear for the first time that assets had been sold, on average, at approximately one-fifth of book value.[665] Even so, the SPA had managed to sell only 20 percent of state assets in three years, and those were the most attractive assets to foreign investors. An independent study published in 1992 predicted that privatization would run out of steam shortly, since most of the attractive assets had already been sold.[666] The good companies would be sold, and the rest would languish until creditors cut off credit.[667] After that, only "dogs" would be left.[668] And in fact that is what happened. For most of 1992, there was little progress at all;[669] the surprisingly high year-end revenue figures came primarily from a few large deals. By October 1992, there were signs that foreign buyers were backing off the privatization process.[670] The remaining state-owned companies—the "dogs"—were hemorrhaging badly, losing approximately $72.3 million in value per month.[671] This fact, rather than the SPA's stated desire to increase domestic participation for economic demographical reasons, may explain why the SPA was shifting in favor of domestic purchasers in 1992.

These disappointing figures do not mean that foreigners were not investing in Hungary. Most foreign investment in Hungary was not through privatization. This was a conscious preference on the part of Western investors. Ownership claims on companies were sometimes unclear, and hidden liabilities militated against buying entire companies. Many Western investors preferred to buy assets or do start-ups to penetrate the Hungarian market without the problems that attend buying entire privatized companies.[672]

Statistics reflect this preference. For example, direct foreign investment in Hungary for the first eight months of 1991 totaled $680 million,[673] whereas privatization revenues for that period were only $127 million, and approximately 20 percent of that figure represented domestic investment.[674] Put together, these data indicate the relative unimportance of privatization, as practiced in Hungary, to overall transformation of a command economy.

1992 and the Shift Toward the Domestic Market

At the end of 1992 therefore, it seemed clear that the transformation of the Hungarian economy would come primarily from the growth of the entrepreneurial class, the continued erosion of traditionally dominant state-owned enterprises, and direct foreign investment other than in purchasing privatized assets. Although privatization would continue to be an important source of revenue, the macroeconomic impact of privatization would probably be overshadowed by other forces as the 1990s progressed. The growth of the private economy was greater than the pace of privatization.[675]

There were, in fact, signs that after 1992 privatization would be less significant economically than it had been that year. By the end of the year, Hungary was lagging behind both Poland and the Czech Republic in the competition for foreign investment.[676] Privatization revenues for 1993 were projected to be approximately $500 million, a 36 percent drop from 1992 levels.[677] Yet, at the same time, the SPA continued to work on creative privatization ideas. In late 1992, the SPA hired Arent Fox and Castellum Ltd. as consultants to help generate new ideas for the disposition of assets that foreigners seemed less and less interested in.[678]

By the end of the year, the SPA's functions had been diminished somewhat by the creation of the State Asset Holding Company, created pursuant to an act of Parliament in the summer of 1992.[679] The State Asset Holding Company would hold the portfolio of state assets being retained by the state, either because the state had reserved a portion of the shares in privatization or because the company would not be privatized. For example, the government took an early position that it would retain a majority stake in the state airline Malev.[680] And it did. After selling a 30 percent stake to the Italian carrier Alitalia and 5 percent to the Italian investment company SIMSET, the state turned its remaining 65 percent stake over to the State Asset Holding Company. The SAHC would retain 51 percent, and the remaining 14 percent would be sold to workers and local investors.[681] With the SAHC engaged in portfolio management, the SPA's function would thereafter be limited to processing companies through privatization.

Despite the efforts of the SPA, more than 95 percent of firms that were state owned in 1988 were still state owned at the end of 1992.[682] Many were clearly dying without hope of privatization, except perhaps through liquidation. The total value of state assets was declining by more than $70 million per month at the end of 1992.[683] Some quality assets of great interest to foreigners (MATAV, the state-owned telephone company, for example) were slated to be privatized soon.[684]

The SPA hoped to use the Budapest Stock Exchange more in the future, following the successful flotations of Pick Szeged and Danubius Hotels in 1992. One official estimated, almost certainly unrealistically, that there would be as many as twenty public-offering privatizations on the BSE in 1993.[685] This was a remarkable turn-around. In October 1991, a leading investment adviser in Budapest said: "Public flotations have practically come to a halt. All transactions are being worked out now on a private placement basis."[686]

Eight

The Budapest Stock Exchange and the Securities Markets of Hungary

One measure of the degree to which a country has developed a private economy is the state of development of its capital markets. But the development of primary and secondary capital markets not only reflects the development of a private economy; it also furthers the growth of that private economy. In Hungary, we see both aspects in the growth of its securities markets. Not only did the rebirth of securities markets coincide with the changeover to a private economy in the late 1980s and early 1990s, but the nurturing of those markets helped advance the general economic reforms. This is particularly true with respect to privatization.

Unfortunately for privatization efforts, the Budapest Stock Exchange was restricted in its development to such an extent that it did not play as great a role as it might have in placing shares of state companies into private hands. To some extent, this was the fault of the SPA. After the IBUSZ scandal of 1990, the SPA was reluctant to use the BSE as a primary privatization mechanism. Only in late 1992, with the Danubius Hotel and Pick Szeged privatizations, did the SPA return to the Budapest Stock Exchange on a large scale. But part of the lack of BSE participation was surely due to the stringent requirements set by the exchange for listing and trading there. As a result, most privatizations were private placements, and most secondary activity took place off the exchange in an informal over-the-counter market.

This chapter chronicles the rise and development of the Budapest Stock Exchange and its participation in the privatization efforts.

The Budapest Stock Exchange and the Securities Markets 111

It also discusses the larger over-the-counter markets that arose concomitantly with the establishment of companies whose shares could be transferred.

But the role of the BSE in a private Hungarian economy is best viewed in historical context. The period in which modern Hungary was without a stock exchange was relatively short—only 42 years. To understand Budapest's history as a center of financial transactions in Europe, we turn to the story of the Budapest Stock Exchange in the pre-Socialist era.

The Old Budapest Commodity and Capital Exchange: 1864-1948[687]

The Budapest Commodity and Capital Exchange[688] officially began operations on January 18, 1864. But the seeds of its development were sown somewhat earlier. The use of intermediary brokers in commercial transactions in Hungary dates back at least to 1840 and the establishment of the "sworn broker" system.[689] But the immediate predecessor of the Commodity and Capital Exchange was the Budapest Wheat Hall, organized by the Pest Lloyd Company in 1854. The Wheat Hall, though less organized than the Commodity and Capital Exchange, served as a more organized commodity trading forum than the traditional weekly fairs that had preceded it.

By an imperial decree of February 1860, exchanges were permitted to be established in large trading centers throughout the Austro-Hungarian Empire. The decree urged only the establishment of commodity exchanges. But after the Hungarian Governor's Council instructed the Pest Chamber of Commerce to work out the rules for an exchange located there, the chamber went forward with a plan for a combined commodity and stock exchange. In 1863 the chamber's plan was approved, and the Pest Lloyd Company (operator of the Wheat Hall) was appointed to set up the exchange.

When it began operations in January 1864, the Budapest Commodity and Capital Exchange was located in the auditorium of the Pest Lloyd Company, where the Liberal party had been meeting for years. The Wheat Hall continued its operations in the same building until 1868, when it was merged into the Commodity and Capital Exchange. In 1873 the Exchange moved to a new building on Valeria

Street. Then in 1904 it moved to an immense and magnificent new fin de siècle building at Szabadsag Square, where it continued to operate until it was closed in 1948.[690]

The founders of the exchange were mainly German-speaking Jews, and the official language of the exchange was German. The perception that it was a Jewish institution would eventually contribute to the closing of the Exchange by Hungarian Nazis, but even in the mid-nineteenth century there was public resentment of the exchange's use of German. In 1893 a new rule was adopted by the exchange that only Hungarian speakers were eligible to sit on the ruling Exchange Council, and gradually Hungarian replaced German as the most prevalent language at the exchange.

When the Budapest Commodity and Capital Exchange was founded, its activity was slow, reflecting the state of the Hungarian economy. In 1864 there were 21 types of securities traded on the exchange, and fluctuations in price came primarily in response to fluctuations in price on the Vienna Exchange, on which the Hungarian shares were also listed. This opened the door for wily traders who could learn about price changes in Vienna before other traders. One such person was Jakab Tauber, who in 1914 described his early primitive excursions into price arbitrage many years before.

> Cables were still slow in those days, and telegrams with news of the Vienna Stock Exchange arrived at the Buda Post Office much sooner than they reached Pest on the other side of the river. So I went up to the Post Office in Buda to pick up cables waiting to be collected, all to the name of my boss. I opened them, and passed them on. We had signs agreed for each day. During the daytime I used a little colored flag, a red umbrella or a white linen sheet, and after dark, when business was still going strong over there [unofficial trading took place in the yard of the exchange after hours], I sometimes used either the red, the white, or the green side of my lantern. My partner, called Szernyi, watched my signs through binoculars from Eotvos Square on the other side. There was in business only a few kinds of papers, so it was easy to do.

The low level of activity did not last long. By 1869, there were 84 securities traded on the exchange, and business on the exchange trended sharply upward for nearly twenty years. There were occasional setbacks caused by external factors such as the outbreak of the Franco-German War and occasional international and state financial crises, but all in all business was good in the Austro-Hungarian Empire in the late nineteenth century. As a result, the exchange prospered with the many new share companies whose stocks were being traded.

Then in October 1894 there were significant slumps in the stock markets of Paris and London, and reverberations were felt in Budapest. That was followed quickly by an 1895 slump in the highly speculative gold mining stocks that had become so popular after gold was discovered in South Africa. On November 9, 1895, the Vienna Stock Market crashed, setting off a slide in the Budapest Commodity Exchange that would last for several years. In 1895, there had been 158 shares traded on the exchange. It did not rise to that level again until 1910. The volume on the exchange in 1905 was only half what it had been in 1895; the 1895 level was not seen again until 1907. There was deep suspicion of speculation in share companies on the part of the investing public, who instead were buying government bonds.[691]

The market did rebound sharply not long after the turn of the century. With the rise of worldwide trade, shareholding companies began playing a leading role in the Hungarian economy. Literally hundreds of such companies were being formed, and many were prospering. In 1905 the Budapest Chamber of Commerce and Industry reported the establishment of 61 industrial companies, 51 financial institutions, and 38 other shareholding companies. As a result, from 1903 to 1907, the stock market was booming.

In 1907 a decline in American railroad construction caused a stock market crisis in Budapest. The enormous railroad construction projects of the late nineteenth and early twentieth centuries had required huge amounts of copper, resulting in high prices for the metal worldwide. Construction had relied upon European capital, however, and that capital became scarce in 1907, leading to a decline in railroad construction in the U.S. and a falling copper price. One shareholding company listed on the Budapest Commodity Exchange that traded in copper collapsed, causing a collapse of its bank. Like dominoes, other firms fell as well, and the decline in the market lasted through 1908. In 1909, however, prosperity came again to the Hungarian economy and the exchange. That period of prosperity ended, however, shortly before war broke out in 1914. Following the closure of the Vienna Stock Exchange in 1914, the Budapest Commodity Exchange closed itself out of fear that Austrian brokers would rush to Budapest to dump shares. Although the exchange remained officially closed until May 1916, unofficially the market was booming outside the exchange as shares of companies pumped up by war-time production soared. By 1917, however, there was a sufficient shortage of coal to force many companies to close. The market slumped from that period to the end of the war.

The immediate postwar period was dominated by inflation (which ran at 904 percent in 1919) and political chaos. As a result, the market went through booms (in 1922, for example) and busts (in 1924, for example). In the period from 1925 to 1927, the government sought to stabilize the economy with a number of measures, including the replacement of the Austrian Crown with the new Hungarian Pengo as the official currency. Again the market boomed.

But by 1928, the tight monetary policy implemented by the government to safeguard its stabilization efforts was having a negative effect on the stock market. Following the collapse of the New York Stock Exchange in October 1929, the Hungarian markets fell into depression with those of the rest of the world. Beginning in 1934, the Hungarian markets began recovering slowly. With some notable setbacks, activity and prices at the exchange moved slowly upward for much of the remainder of the decade.

The threat of war in Europe and the anti-Jewish laws of 1938 and 1939 caused the market to slump again. There was some strengthening when the Hungarian economy went on a war footing; indeed, the Budapest Commodity Exchange experienced so much speculation that in early 1943, maximum daily price rises of 1 percent were imposed on shares there, and on February 11, 1943, prices were fixed.

In October 1944, however, an ultraviolent fascist regime took power in Hungary. The government forced the president of the exchange to resign. Two months later the battle of Budapest began, and the market was closed again. It reopened shortly after the war ended. But in 1948, the Hungarian Socialist Workers party consolidated its power and declared the activity of the exchange to be illegal. After 84 turbulent years, the Budapest Commodity and Capital Exchange was dead. From that time until the early 1980s, there was no securities market in Hungary, let alone an organized exchange.

Trading Activities Prior to the Formation of the Budapest Stock Exchange

Ironically, it was the Communist regime of Janos Kadar that sowed the seeds of a new stock exchange in Hungary. The growth of what eventually became the Budapest Stock Exchange began modestly. In 1982 the Hungarian Parliament (still very much under the control of the Hungarian Socialist Workers party) passed legislation permitting companies to issue debt and equity securities.[692] At first, individuals were not permitted to purchase either type, although they were allowed to buy government bonds.[693] That changed quickly; by the summer of 1983 individuals were buying enterprise bonds too.[694] Individual share purchases were not legal until passage of legislation in 1988.

Impetus for the legislation came from quarters in the government that were exploring the envelope of free market thinking in the Socialist system. According to Peter Akos Bod, a leading MDF economist and one-time head of the Hungarian National Bank, the goal was more efficient resource allocation. Unless companies could issue shares or bonds, profits had to be invested internally. But there arose a perception that if those profits could be more efficiently invested in another firm, legislation should permit such activity.[695] As Janos Fekete, then deputy president of the Hungarian National Bank, explained in 1983: "There are a lot of people with money who do not use it, and there are others who need money. Why not let them change positions?"[696] Communist party chief Janos Kadar, however, insisted these were "not capitalist methods but socialist methods of a socialist society."[697]

Following passage of that legislation, a bond market developed rapidly. In March 1983, the State Development Bank underwrote the first nongovernment bond issue in postwar Hungarian history, issuing on behalf of the Hungarian State Oil and Gas Trust approximately $5 million worth of bonds at an interest rate of 11.5 percent for ten years. The issue was a huge success. It was vastly oversubscribed, in part because with inflation stable at between 7 and 8 percent, the bonds offered an attractive real rate of interest. Internal investment could not often guarantee such a return on investment.[698]

Very quickly the State Development Bank developed a booming business in both underwriting and brokering bonds. Terms were sometimes creative. In late 1983, the Hungarian Post Office sold 6 percent bonds to residents of southeast Hungary by promising purchasers installation of a telephone within three years; ordinary customers had to wait as long as five years.[699] The issue was oversubscribed. A similar scene occurred two years later in Debrecen.

When the Post Office announced that it would begin selling bonds that included a shorter wait for a telephone, people started queueing up the night before the sale was to begin. Those who bought bonds jumped ahead of those 12,000 who were already on a waiting list for phones. There was apparently little grumbling by the 12,000 already on the waiting list, though; they figured they had little chance of getting a phone anyway.[700]

By early 1985, a total of 28 bond issues had taken place, with an aggregate value of approximately $30 million.[701] And the individual at the State Development Bank responsible for the development of that business, Zsigmond Jarai, can be aptly described as the grandfather of the Hungarian securities markets.

Jarai was only 32 years old when he helped the State Oil and Gas Trust make the first bond issue in 1983. He immediately became the head of the State Development Bank's burgeoning bond department and nearly single-handedly developed an ever-busy secondary market for his bonds, starting in 1984.[702] In 1985 he set up a trading room of sorts at the State Development Bank to facilitate trading. The room consisted of a table with a computer screen that showed the price and number of bonds available. Investors would crowd around to do business with Jarai.[703] Turnover in 1985 averaged approximately $40,000 to $60,000 per day.[704] Even then Jarai talked of developing a market for shares within the bank.[705]

In 1985, Jarai's bond business really took off. By the end of the year, 80 issues (worth approximately $83 million[706]) were in circulation, and Jarai was making a market in all of them. More than 60 percent of the bonds were bought by individuals, who at first showed a strong preference for short-term bonds. But when they realized the strength of Jarai's secondary market, they became more willing to accept longer terms.[707]

By July 1986, there had been 120 issues of bonds, with a total value of approximately $150 million. Approximately 30,000 Hungarians had invested in them, and Jarai calculated that the number of customers was doubling approximately every six months.[708] When in 1987 the State Development Bank's business was spun off to the new private Budapest Bank Rt.,[709] Jarai went too, taking the bond business with him. It was while he was at Budapest Bank that he helped organize the group that eventually became the Budapest Stock Exchange.

The need to organize the secondary bond market arose from its size. By early 1988, there were approximately $570 million worth of bonds in circulation. And after the banking reform of 1987, there were several commercial banks that were trading in bonds, although Budapest Bank and Zsigmond Jarai controlled approximately 90 percent of the action.[710] Arbitrageurs emerged, buying bonds at one bank and selling them at another. As Jarai told a reporter in early 1988: "Every commercial bank would like to create its own market. But we must work together, we must regulate new issues and we must regulate the secondary market."[711]

Establishment and Organization of the New BSE

Thus, on January 12, 1988, the National Bank of Hungary, the Hungarian Chamber of Commerce, and 22 financial institutions signed an agreement for a regulatory framework to organize the bond market: the Agreement on Trading in Securities. There was no real stock market yet, since at that time only enterprises could buy common stock. But the banks that signed the regulatory accord agreed to meet once a week for a "dealers' day" of interbank trading of bonds. They further agreed to establish a central information center and transmit to it on a daily basis information concerning price and volume for all bonds being traded.[712]

Pursuant to the agreement, the dealers' day was run by a Secretariat for Trade in Securities, headed by Ilona Hardy, who would later become the first managing director of the Budapest Stock Exchange.[713] The dealers' days were held in a windowless second-floor office in the modern International Trade Center in central Budapest, which functioned as a "trading floor." In fact, there was often a good bit of quiet around the table where the participants sat. Sometimes several minutes would go by between trades. Only 1 or 2 percent of all trading activity took place at the meetings; most took place during the week in the banks' offices.[714] Still, by the summer of 1988, the participants were making plans for the formal establishment of the Budapest Stock Exchange, and they aimed for an opening date of January 1, 1989. In preparation, on July 26, 1988, the signatories to the Agreement on Trading in Securities elected a governing council for the stock exchange and elected Zsigmond Jarai its first chairman. Its first task was to draw up a charter for a new stock exchange.[715]

Passage of legislation in 1988 authorizing the formation of companies limited by shares and permitting private persons to own such shares gave further impetus to the fledgling exchange. Many new economic associations being formed were companies limited by shares, and privatization was making available some shares of former state-owned enterprises.[716] But formal establishment of a securities exchange awaited passage of authorizing legislation. That legislation was being drafted in 1989, but it was not passed until January 1990. So planning went on throughout 1988 and 1989. The Stock Exchange Council obtained a loan from the World Bank to buy data processing equipment and established formal ties with the London Stock Exchange to receive technical assistance.[717]

In the summer of 1989, Zsigmond Jarai left Budapest Bank to become Deputy Minister of Finance, and he was placed in charge of banking and securities oversight. As a result, direction of the drafting of legislation authorizing the establishment of the Budapest Stock Exchange fell to him. He decided that such a statute should not only authorize the exchange but regulate the securities markets generally to prevent fraud and market manipulation.[718] In anticipation of the legislation, the Stock Exchange Council established a Securities Trading Committee (headed by Ilona Hardy) to handle the technical aspects of setting up the exchange and a draft set of membership requirements for those who wanted to obtain memberships in the planned Budapest Stock Exchange.[719] When the necessary legislation was finally passed in January 1990,[720] the individuals driving the exchange were nearly ready to open it for business.[721] Stocks were now being traded alongside bonds at the trading meetings, although approximately 80 percent of all securities transactions were taking place away from the trading meetings.[722]

There was no sense of urgency, however. Trading at what people were now calling a securities exchange (but in reality just a more computerized dealers' day site) was slow in early 1990. One reporter who in February visited the room in which trading took place described the atmosphere: "With a dozen personal computers placed neatly around a square table in a stuffy, windowless room, Hungary's stock exchange seems more like a student pilot project than a functioning financial market."[723] The pattern seen in privatization efforts—lofty expectations never met—was being replicated at the

Budapest Stock Exchange. Even before the exchange officially opened in June 1990, its backers were expressing surprise and disappointment at how slowly activity was growing at the pre-exchange trading meetings. In October 1989 the meetings were taking place twice per week instead of only once, and the following month the meetings began taking place three times per week.[724] Still, volume was anemic. Average daily turnover was only approximately $80,000 in February 1990. Only about one-fifth of that volume represented share trading; the rest was bond trading.[725]

The reasons for the lackluster pace of activity were many. A lot of trading was still taking place at the banks' offices, and such over-the-counter activity would continue to dominate trading volume well into the 1990s. The bond market had collapsed in 1988 following a surge in inflation, and the slow pace of privatization meant the number of shares available for trading was small. What shares there were traded extremely thinly. Inflation also made the general population hesitant to buy equities (whose returns were uncertain and long-term) or bonds (which locked them into an interest rate for several years); instead those who had any spare money to save purchased three-month or six-month certificates of deposit.

But perhaps the most important reason for the failure of the stock and bond markets to catch on was lack of confidence and understanding on the part of the people. After 40 years of Socialist indoctrination, most were not eager to participate in what seemed, at best, a gambling game and, at worst, immoral. Official ideology deemed a stock exchange a "parasitic institution where speculative transactions are carried out" and "where it is possible to obtain incomes without the actual performance of work."[726]

And even if they had wanted to invest, most Hungarians had no idea what stock and bonds were, let alone how to judge which were good investments and which were poor. As one broker put it: "We never learned about these things in the Hungarian education system. We're like children on these matters. We have to be taught the language."[727] As a result, money went into real estate, consumer goods and other items of consumption, and hard currency.[728]

Still, work continued on both a new physical site and an organizational structure for the exchange. The authorizing legislation, Act VI of 1990 on Securities and the Stock Exchange, had mandated a certain governance structure for the exchange, and the Stock Exchange Council was busy hammering out details throughout the winter and spring of 1990. A general founding meeting of all prospective members of the new exchange took place on June 19, 1990, at which the members ratified the formation documents that had been prepared by those organizing the exchange. Finally, with the organizational and financial structure in place, the morning of June 21, 1990, was set as the time for the official opening of the newly chartered Budapest Stock Exchange.[729] Unfortunately the new trading facilities on Deak Street, a few blocks away from the International Trade Center, were not yet ready,[730] so the grand opening had to take place in the old trading room.

The number of distinguished guests on hand for the historic moment outnumbered the brokers by a wide margin. Those on hand included representatives of the American Securities Exchange Commission, a law firm that was trying to claim a spot as official consultant to the exchange, and a group of American law professors who happened to be in Budapest at the time. There were several solemn speeches made about the importance of what was about to happen, and the sense of anticipation was palpable. As the exact moment for the commencement of trading drew near, the room became quiet.

Finally, a bell rang and there was much applause. Then nothing happened. No one had thought to line up a few big trades to get the ball rolling. Instead, there were no trades for nearly thirty minutes as the gathered brokers and dignitaries stood around looking sheepish.[731]

Eventually trading did begin and took place every business day thereafter, although activity never met expectations, despite occasional bursts of excitement. Pursuant to the Securities Act, the Budapest Stock Exchange was founded as a nonprofit independent legal entity not connected to the state.[732] Also pursuant to law, the exchange was created as a self-regulatory organization on the American model. It was required to establish its own governance system, and rules and regulations were to be developed internally, subject to governmental oversight.[733]

There were 41 founding members of the exchange, and they included nearly every financial institution in Budapest.[734] The initial capitalization of the exchange amounted to the forint equivalent of approximately $2.86 million, most of it in cash, and most of it contributed by one state organ or another. The following table shows the breakdown of the exchange's initial contributions:

Cash	$1,930,573
41 members @ $40,000 each: $1,640,000	
Hungarian National Bank: $290,573	
Leasehold @ 5 Deak Street	$ 821,333
Equipment	$ 109,413
Total	$2,861,319

    The leasehold was contributed by the State Development Institute, a state organization that was the nominal owner of the building where the exchange would be housed. The lease was for an indefinite (i.e., permanent) term, making in effect a contribution of the building in fee simple. The equipment contributed by the Hungarian National Bank included items such as copiers, printers, computers, and fax machines. Although the cash came primarily from the founding members, most of those founding members were owned, either in whole or in part, by the state.

    Organizationally, the initial charter of the Budapest Stock Exchange called for a multitiered governance structure. The "supreme organ" was said to be the general meeting of all the members.[735] Such a meeting had to be held at least annually in the first three months of the year, although the charter provided for the calling of special meetings. The primary function of the general meeting was to elect officers of the other governance organs.[736]

    The most important of those other organs was the Stock Exchange Council. Elected annually by the general meeting, the council acted as a sort of board of directors made up of between five and eleven members. Its primary function was the establishment and amendment of stock exchange rules. Its appointee acted as chairman of the annual general meeting of members, and the council chose its own chairman.[737] As head of the highest policy-making body of the exchange, the chairman of the Stock Exchange Council occupied the highest office of the exchange.

    The Supervisory Committee was established by the charter as an executive body charged with controlling the finances of the exchange and overseeing day-to-day operations, including trading activities. Its members, elected by the general meeting of members, were also responsible for ensuring compliance with the rules of the council.[738]

    The Securities Trading Committee's function was to execute the operations of the exchange on a day-to-day basis. Headed by the managing director of the exchange, the Trading Committee was "obliged to organise trading on the Stock Exchange, to control the Stock Exchange, to put the decisions of the General Meeting and the Stock Exchange Council into execution, to disclose information on the Stock Exchange, and to carry out the economic activities of the Stock Exchange."[739] Because the managing directorship represented the highest functionary position at the exchange, its holder was also the person of highest visibility at the exchange.

Finally, the charter called for the existence of two more specialized bodies. The Ethics Committee was charged with rendering decisions on alleged ethical breaches by members.[740] The Arbitration Court was established to hear disputes among exchange members.[741]

This organizational structure, labeled "excessively bureaucratic"[742] by more than one participant, clearly contemplated a much more sophisticated and busy institution than the Budapest Stock Exchange was in the summer of 1990. The founders were evidently anticipating rapid growth, and they wanted in place an organization prepared to oversee much more activity than that going on at the time of founding.

Activities of the Exchange, On the Exchange and Off the Exchange

In the fall of 1990, the Budapest Stock Exchange was able to move into its new facilities in an old stone office building on Deak Street. Trading took place in a room set aside for that purpose, although even then plans were underway for the creation of a large, modern, computerized trading floor. The floor was open for trading Monday through Friday from 10:30 A.M. until noon. When I first visited the exchange's new facilities and expressed surprise that trading lasted 90 minutes per day, an official of the exchange leaned forward, put his hand on my shoulder and confided to me: "Yes, I know. It's far too much time for our limited market to be open. But we want to be prepared when activity expands."[743]

By the end of 1992, trading was taking place in a beautiful and modern new room that retained architectural flourishes of the original building, such as a ceiling of glass and arches perhaps 25 feet above the floor. The room was approximately 80 feet long and 40 feet wide with a raised dais on one side for the person overseeing trading. On the other side of the room was a double row of desks — one for each broker — each with its own telephone. At the end of the room sat a bank of approximately two dozen IBM-compatible personal computers.

Trading was by computer for those shares with good liquidity. For those without substantial activity, trading took place by open outcry. Once confirmed by the parties to the transaction, the trade ticket was turned in to the staff for processing by the clearance and settlement system.

Spectators could observe activity from behind a glass wall at one end of the trading room, but access to the floor itself was restricted to registered brokers by security personnel. Communication between brokers and spectators was strictly prohibited except by phone. Nonetheless, several speculators could be seen on any given day trying surreptitiously to signal instructions to a broker on the floor.

The visitors' lobby behind the glass was a modern, well-lit, marble-floored area with a bank of six interactive color computer screens imbedded in the wall. The computers served to educate visitors in the fundamentals of investment. After choosing the language desired (Hungarian, German, or English), the visitor could get instruction on topics such as "What Is a Free Market?" or "What Is a Stock Exchange?" or "What Is Capital?" or "What Is the Difference Between Dividends and Interest?"

The actual trading was done by licensed brokers or dealers, who were required to pass an examination administered by the State Securities Supervisor. At the end of 1992, there were 48 such brokers and dealers. Forty were dealers, meaning they could trade either for their own accounts or for customers; the other 8 held only broker's licenses, meaning they could trade only for the accounts of customers.[744]

After moving into the new quarters, one of the first tasks of the Stock Exchange Council was the establishment of a formal set of rules. It adopted the Rules of the Budapest Stock Exchange on October 5, 1990, although the rules were not approved by the State Securities Supervisor until December. Until then, the exchange operated under a combination of established past practices and the fiat of the managing director, Ilona Hardy.

The exchange permitted two categories of securities to be traded: "listed securities" and "traded securities." The former category was more difficult to attain, but was thought by the BSE to provide a badge of prestige to the security, particularly with foreigners.[745] The following comparison shows the differences between the two categories.

TABLE 8-1: REQUIREMENTS FOR LISTED AND TRADED SHARES ON THE BSE[746]

	Listed	Traded
Capital required	200 million HuF	100 million HuF
Percentage that must be publicly held	25	10
Financial history required	3 years audited	None
May be traded where?	Only on Exchange	Exchange or over-the-counter

Each of the categories also has other requirements regarding disclosure of information and the like,[747] but the above table shows the important differences between the two. Given the greater burdens placed on the listed securities, it should be no surprise that most securities in which the BSE has activity are of the traded variety. But even the requirements for the traded category have proven too onerous for some issues. When the market first opened, there were shares of approximately 60 companies traded on the exchange, but all but a handful left when the traded category requirements came into effect on December 31, 1990. Today in Hungary, most estimates put the percentage of securities transactions taking place away from the BSE at approximately 90 percent.[748]

When the Budapest Stock Exchange first opened, there were dozens of shares traded under a grandfather provision that gave the issuers several months to comply with the trading requirements. Many dropped out rather than meet the requirements. At the end of 1992, there were seven shares in the listed category and 17 more in the traded category. The following tables show the shares in both categories together with volume and price information for 1991 and 1992. The tables reflect both the anemic volume on the exchange and the downward trend of most share prices. The tables also illustrate

TABLE 8-2: 1991 VOLUME AND PRICE DATA FOR SHARES ON THE BUDAPEST STOCK EXCHANGE, IN U.S. DOLLARS[749]

Company	Yearly Volume	Value per Share			
		High	Low	Open	Close
Listed Shares:					
Dunaholding[750]	5,356,533	533	293	467	387
Fotex[751]	15,332,133	3.61	2.47	2.91	2.73
IBUSZ[752]	1,664,080	80	32	63	40
Konzum[753]	3,526,040	39	9	30	9
Styl[754]	6,143,947	49	41	41	46
Sztrada Skala[755]	691,733	362	173	268	245
Zalakeramia[756]	1,097,466	23	20	21	22
Traded Shares:					
Agrimpex[757]	67,466	469	180	467	293
Bonbon Hemingway[758]	167,440	37	24	35	25
Budaflax[759]	43,173	22	12	22	12
Elso Magyar Szovetkezeit Sorgyar[760]	931,573	40	28	29	33
Garagent[761]	38,400	220	212	221	220
Hungagent[762]	20,400	36	32	32	35
Kontrax Irodatechnika[763]	981,600	347	309	309	345
Kontrax Telekom[764]	915,467	336	308	308	336
Muszi[765]	1,266,133	252	147	215	200
Nitroil[766]	185,867	293	205	205	240
Novotrade[767]	126,906	59	13	59	15
Skala-Coop[768]	2,733,067	431	160	293	173
Terraholding[769]	82,733	22	12	22	12
Total Volume:	40,372,157				

TABLE 8-3: 1992 VOLUME AND PRICE DATA FOR SHARES ON THE BUDAPEST STOCK EXCHANGE, IN U.S. DOLLARS[770]

Company	Yearly Volume	Value per Share			
		High	Low	Open	Close
Listed Shares:					
Danubius[771]	2,133	13	13	13	13
Dunaholding	9,480,000	413	200	387	216
Fotex	33,506,667	3.84	2.47	2.73	3.79
IBUSZ	11,114,667	54	25	41	47
Konzum	2,486,933	11	4	9	5
Styl	4,654,000	81	44	48	73
Sztrada Skala	307,867	240	67	240	73
Zalakeramia	295,067	23	16	21	16
Traded Shares:					
Agrimpex	199,333	400	293	317	293
Bonbon Hemingway	189,867	25	17	19	17
Budaflax	291,867	13	8	13	8
Elso Magyar Szovetkezeit Sorgyar	467,933	37	15	33	15
Fonix[772]	31,333	18	16	21	16
Garagent	661,600	220	203	217	207
Hungagent	304,400	45	15	32	15
Kontrax Irodatechnika	706,133	347	200	347	300
Kontrax Telekom	893,333	336	189	336	240
Muszi	107,200	193	36	193	36
Nitroil	476,267	240	200	240	200
Novotrade	476,767	14	7	14	8
Pick Szeged[773]	1,126,133	17	16	16	16
Skala-Coop	4,031,600	167	95	164	132
Terraholding	3,467	12	4	12	5
Total Volume	71,329,300				

the dominance of certain shares in the overall volume figures; in 1991, for example, Fotex accounted for approximately 38 percent of the entire BSE volume.

The figures shown in the above two tables do not include important classes of securities that accounted for substantial amounts of activity on the exchange in 1991 and 1992. Strictly speaking, the Budapest Stock Exchange was from the beginning more than a stock exchange; other types of securities were traded there as well. Indeed, the Hungarian name for the exchange, "Budapesti Ertektozsde," is more accurately translated "Budapest Securities Exchange." The exchange's roots were in the bond market, and bonds continued to represent a significant part of the exchange's volume. Debt securities, primarily in the form of short-term government obligations, accounted for approximately $349 million in 1992, a substantial majority of the total volume on the exchange.[774]

Originally, by exchange rule, there were five categories of securities that could be traded on the exchange in the early 1990s: shares, bonds, government debt securities, options, and futures.[775] Most of the share activity took place in the trading of just a few of the best known shares and, increasingly as the share market became less attractive to investors, in the trading of government bonds and bills. Compensation coupons traded on the BSE by the end of 1992 and were also traded in an extremely informal over-the-counter market. Throughout 1992 one could find advertisements in several of the Budapest newspapers for the purchase or sale of compensation coupons.

The above tables also do not include information for investment fund activity, which accounted for approximately $396,800 in volume that year, or for options, futures, or compensation certificates, which together accounted for several millions of dollars in activity in 1992.[776] The following table shows the breakdown of all securities activity on the exchange for 1992.

The small investors in Hungary received an important boost in June 1992 with the establishment and trading on the BSE of CA Investment Fund, a closed-end mutual fund. With meager individual savings, it was difficult for average citizens in Hungary to buy into the share market without betting too great a percentage of their savings on the fortunes of a very small group of companies. In addition, although brokerage firms adamantly disavowed the practice, several credible sources reported to me in 1992 that brokers routinely

TABLE 8-4: BREAKDOWN OF ACTIVITY
ON THE BUDAPEST STOCK EXCHANGE, 1992,
IN U.S. DOLLARS[777]

Type of Security	Dollar Volume	% of Total BSE Activity
Listed Shares	61,847,334	14.52
Traded Shares	9,481,966	2.23
Investment Funds	396,800	.09
Debt Securities	348,910,000	81.89
Options and Futures	2,574,106	.60
Compensation Certificates	2,848,000	.67
Total Volume	426,058,206	

turned away business of the small investor as being too time consuming for the commissions involved. Thus, when legislation permitting the establishment of investment funds was passed in 1991,[778] a new avenue for pooling small investor resources opened. Creditanstalt became the first issuer of shares in such a fund when it opened its CA Investment Fund in 1992, which began trading on the BSE in May of that year as a listed company.

In late 1991, the SPA was reportedly considering establishing more than one investment fund into which it would pour state property and shares which it would sell or make available for compensation coupons.[779] By the end of 1992, no such investment funds had been established, though.

A separate organization was established to facilitate trading in commodities. This Budapest Commodities Exchange maintained a relationship with the Budapest Stock Exchange, but the former was more commercial and less officially sanctioned than the latter. Whereas the BSE was a self-regulatory organization explicitly provided for by law, the BCE was simply a Kft. In late 1992, there were reports circulating that the BCE was considering opening a financial futures exchange as early as the first quarter of 1993.[780] Exchange officials foresaw a confrontation between the BCE and the BSE over trading of derivatives such as warrants, options, and financial futures, a confrontation mirroring that between the American SEC and CFTC.

Clearance and settlement of transactions on the exchange took place at the exchange itself through 1992 pursuant to a set of settlement regulations adopted by the Stock Exchange Council in October 1990.[781] Many connected with the exchange spoke of plans to create a separate institution to handle clearance and settlement, but as with so many plans in the development of Hungary's private economy, they remained only plans through the end of 1992. One of the problems for settlement purposes was the fixation, both in law and in the mind-set of Hungarian traders, with the physical share certficate. The Hungarian term for security, *ertekpapir*, means literally "value paper," and the term for share, *reszveny*, also connotes the physical paper representing the ownership interest. Thus, securities in general and shares in particular were understood not as intangible bundles of legal rights represented by the certificates, but rather as the certificates themselves. The Hungarian Civil Code also required physical delivery of the security to effectuate a transaction. This made extremely difficult the introduction of computerized trading by book entry. Hungarians simply had difficulty grasping the concept of stock certificates as being mere evidence of ownership. Headway was being made by the end of 1992, and there were movements toward amending the Civil Code.[782]

A quiet scandal took place at the exchange in the fall of 1992 when the managing director of the BSE, Ilona Hardy, suddenly resigned. Although she was reported in the press as having left for personal reasons and to pursue other activities, insiders were telling a slightly different story: she had threatened to resign one too many times over policy differences, and finally her bluff was called by the Stock Exchange Council. The BSE lost very little with her departure, however, because by all accounts her replacement, Lotfi Farbod, was extremely competent. Farbod was young (29 years old) when he took the position, but many top financiers and government officials were young at that time. He was born in Tehran, Iran, and moved with his parents to Hungary when he was a teenager. Educated at the agricultural university in Debrecen, Farbod first worked for Muszi, the agricultural firm whose shares were traded on the BSE as early as 1991. Shortly after the exchange was opened, he became the manager of the Listings Department and eventually a member of the Trading Committee. In October 1992, he succeeded Ilona Hardy as managing director.[783]

Despite the optimism surrounding the opening of the exchange, the first few minutes of its existence turned out to be a metaphor for its operation as a whole through 1992. Except for the trading in short term government debt securities, volume was always anemic, and the trading very thin for all but a few of the approximately one dozen securities traded. An index was established in 1991 to plot the aggregate rise and fall of the market, with the opening of the index pegged at 1000. By the end of 1992, the index was in the 700s and still falling. The following table demonstrates the lackluster performance of the Budapest Stock Exchange in terms of value for the first years of its operation.

TABLE 8-5: THE BUDAPEST STOCK EXCHANGE INDEX, MONTHLY OPENING LEVELS, 1991-1992[784]

1991	Jan.	1000.00
	Feb.	1075.58
	Mar.	1107.90
	Apr.	1196.37
	May	1164.01
	Jun.	1099.48
	Jul.	1050.30
	Aug.	978.35
	Sep.	924.13
	Oct.	856.81
	Nov.	811.25
	Dec.	784.92
1992	Jan.	808.33
	Feb.	813.03
	Mar.	813.80
	Apr.	877.29
	May	873.09
	Jun.	865.82
	Jul.	862.75
	Aug.	830.65
	Sep.	825.46
	Oct.	835.03
	Nov.	812.60
	Dec.	765.18

Although the BSE index lost approximately 24 percent of its nominal value in the first two years of its existence, the nominal figures paint a misleadingly rosy picture of values on the exchange. With inflation running at between 20 and 30 percent in the period 1990-1992, the real values on the exchange, adjusted for inflation, were slipping precipitously.

The Budapest Stock Exchange and the Securities Markets 131

The disappointing progress at the Budapest Stock Exchange was both a cause and effect of the failure on the part of the SPA to privatize state companies by floating their shares on the BSE. The failure on the part of the SPA to make more than limited use of the BSE resulted in the market's remaining very small. And the small size of the market made it an unattractive forum for privatizations.

Still, a few of the most successful or highly publicized privatizations took place on the exchange, including IBUSZ, Danubius Hotels, and Pick Szeged, all discussed in previous chapters. The exchange will undoubtedly play a greater role in future privatizations as it achieves its critical mass. And, indeed, 1993 turned out to be a turnaround year for the exchange in some ways. By the end of the year, the BSE index stood near 1,200, up almost 43 percent for the year and outstripping inflation by a long way.[785] In addition, the exchange added in 1993 two new shares: those of Zwack Unicum (the large liqueur manufacturer) in the listed category and those of Csemege Julius Meinl (grocery shops) in the traded category. There were also five new closed-end investment funds added to the exchange in 1993, which increased the investment opportunities for the small investor.

Still, as a share market, the BSE continued to slide in importance in 1993. In the first five months of that year, government debt obligations accounted for a whopping 96 percent of the volume on the exchange, whereas equities accounted for only 2 percent.[786] This was a particulary disturbing trend in light of the fact that in the period from June 1990 through May 1993, equities accounted for 22.3 percent of the total turnover on the exchange. Thus, although the equity turnover increased some 78 percent from 1991 to 1992, as a percentage of the activity on the exchange, it was becoming marginal as 1993 passed.

Obviously the usefulness of the Budapest Stock Exchange for privatizations is partially within the control of the BSE. By relaxing some of its requirements and doing away with the dubious distinction between listed and traded securities, it could increase its volume enough to make it a more attractive place to offer privatization shares, either by the SPA or in secondary offerings by the winning bidders themselves. Without taking dramatic steps to open its facilities to more participants in the capital markets of Hungary, the BSE risked becoming an irrelevancy, quickly passed by a burgeoning OTC market trading on computers throughout Hungary and in financial centers throughout the world.

Nine

The Ongoing Debates — A Conceptual Framework for Thinking About Privatization Efforts

The Hungarian efforts at privatization, while unique in their circumstances, offer lessons for large scale privatizations generally. Before addressing those lessons directly, however, this chapter creates a conceptual framework for drawing conclusions from the Hungarian experience. Before getting to the answers provided by Hungary's experience with privatization, we must formulate the questions and examine them in light of the Hungarian experience. First, we take up the questions that have formed the elements of debate in Hungary.

How Fast to Go?

There are two important aspects to an analysis of the appropriate pace of privatization in a country such as Hungary. One is the general policy question: at what speed should a privatization program be carried out? The other aspect is more pragmatic and involves an examination of the costs and benefits of the pace (i.e., moderately slow) chosen by the Hungarians. This section takes them up in turn.

The Pace Debate

The question of how fast to proceed partially (but not entirely) devolves into the question of what method or methods of privatization are preferrable. A massive giveaway of property can be accomplished relatively quickly, whereas sales at high prices take much longer. Thus the debate over pace in Hungary was largely about method. But it was not entirely about method. This is because it may not be prudent to move at the most rapid pace a given method would allow. We see in examining the terms of debate in the late 1980s and early 1990s that commentators and policymakers were concerned not only with method but also with moving at a pace that was sustainable politically and economically over the long term.

During the national election campaign that led to the MDF forming a coalition government in the spring of 1990, the debate concerning pace was partly one of rationality vs. speed.[787] Spontaneous privatizations were occurring unchecked. The MDF campaigned on a platform of slower, more controlled privatization that would divest 50 percent of state property in four years,[788] and the party protested against the "cut-price sell off of the national heritage."[789]

The main opposition groups, SDS and Fidesz, campaigned for a more rapid privatization with the privatization process itself privatized. The opposition groups argued that any government role should be limited to that of watchdog and that the problems that would attend lightly controlled privatization were acceptable, given the primary need to put the means of production into private hands.[790] In addition, the opposition predicted that tax revenues from increased economic vitality would boost budgetary shortfalls and help pay off the foreign debt.[791] As Dr. Morton Tardos, consultant for the SDS, stated the party's position just before the elections: "What we're proposing is the rapid privatization of the entire economy. It's something unprecedented in size, speed, and daring."[792] Speed was a lot more important than price to the opposition groups.[793]

While the problematic and uncontrolled spontaneous privatizations were taking place before the 1990 elections, many argued that Hungary was moving too quickly. Advocates of rapid privatization argued, however, that three benefits would attend moving quickly: a middle class would arise, ensuring stability of the new democratic political system; revenue would be generated to help pay the foreign debt, so long as the state ensured that it (and not the nomenclatura) received the proceeds of privatization; and the economy would be reinvigorated by private economic actors.[794]

This last point was the one emphasized most in the early debates. As a Bulgarian economist looking at his own country noted:

"We need a quick move to a market economy. And for such a change to succeed we need most of our firms to be privately owned."[795] Gyorgy Matolcsy, the Hungarian privatization specialist, explained the mechanism whereby private ownership would stir economic growth:

> Privatized firms will invest more readily in new technology and equipment, they will bring in better know-how and management skills and will organize themselves more efficiently than their state-controlled predecessors. They'll encourage entrepreneurship and incentive, generate profit and get the economy moving.[796]

Or, as a writer in the *Economist* suggested at about the same time:

> Ownership change is the key: real improvement [in the economy] will not come about until genuine private owners take responsibility for the inefficient and nomenclatura-dominated economy.... They must privatise enough, and fast enough, to make the difference felt quickly.[797]

Not everyone shared the sanguine belief in the miracles that would attend rapid privatization. One commentator writing in mid-1990 suggested that privatization would not necessarily lead to marketization; rather, economic effects of privatization would depend on how the process evolved.[798] David Newberry has argued that studies of the United Kingdom's privatizations show that most of the economic efficiencies came in the run-up to privatization. To him that suggested that the efficiencies resulted from the government policies put in place by the Thatcher government before the actual privatizations. From this he concluded that creating proper economic conditions is more important for Hungary than is ownership structure.[799]

Still, the economic data from Hungary's Central Statistical Office seemed to demonstrate the economic superiority of private businesses over state-owned enterprises. According to figures released in 1991, joint venture companies outperformed others in Hungary in a number of respects. They produced a 62 percent higher per capita contribution to GDP, they showed a 25 percent greater return on sales, and they showed a 27 percent greater return on assets.[800]

Outright critics of rapid privatization countered the arguments in favor of rapid privatizations with both principled and pragmatic objections. Domestic Hungarian individuals had little to invest, they said. According to one banker, the new bourgeoisie was "fragile and very thin."⁸⁰¹ In addition, Hungarian institutions lacked funds for investment, and all potential Hungarian investors lacked a sufficient familiarity with securities.⁸⁰² All this pointed to nearly complete foreign domination of privatization sales if those sales were made too rapidly. One observer estimated in the fall of 1990 that, even if credit were available, Hungarians could purchase only about 10 percent of state property even if they were so inclined.⁸⁰³

Given the populist/nationalist character of the MDF and its coalition partners (particularly the Independent Smallholders party), one suspects that this was their fundamental objection to rapid privatization sales: all the property would go cheaply to foreigners. Much of the rhetoric of the MDF suggested this. These groups clearly feared a public backlash if sales to foreigners were made too pervasively and rapidly. As one official put it in early 1990, "Many people feel that the country is being sold off and that they are being cheated."⁸⁰⁴

But many Hungarians saw no alternative to selling to Westerners. Andras Simor, a leading brokerage firm executive in Budapest, told a reporter in 1991 that "large companies cannot be sold to widows and orphans but to people who can manage them. There is no option but to sell them into foreign hands. For the majority of companies we need active shareholders."⁸⁰⁵

Those in favor of a slow pace still argued that if a "fair" (as opposed to market) price were to be obtained, proper valuation would have to take place, and this would take time. Auditors could assess profitability and cash flow, but it was difficult to assess the degree to which central planning and fixed prices distorted the figures. Not only was it going to be difficult to assess how individual firms would fare in a competitive economy, it was also difficult to foresee what general economic conditions lay in the future.⁸⁰⁶ As one wag noted in 1990, "Unfortunately it's very hard to say what's cheap and what's expensive when there isn't a handy yardstick."⁸⁰⁷

Others suggested that Hungary would have to trade quality of price for speed: "In [Central and Eastern European] economies with vastly overextended public sectors, it might in many cases be inevitable to trade quality, i.e. optimal price and ownership structure, for increased speed."⁸⁰⁸ Privatizing quickly seemed to many to be inconsistent with selling at a high price.⁸⁰⁹

Because of the difficulties of selling assets quickly, some suggested that if Hungary wanted to move quickly, it would have to give assets away. Quick sales without proper market mechanisms would simply result in entrenchment of economic power in the hands of the nomenclatura, ossifying existing economic hierarchies.[810] On the other hand, one advantage of moving slowly was that property could be sold rather than given away, and this would result in a decrease in Hungary's foreign debt burden.[811] That debt amounted to approximately $20,000 for every person in Hungary in 1990, the highest per capita foreign debt load in Eastern Europe,[812] so possible repayment in the course of privatization was an important consideration.

Respected financial voices also cautioned against a hurried pace. Janos Kornai, elder statesman of Hungarian economists, was among them: "By all means start the process now. But take it easy. It's important to remember that companies can only be sold if there are sufficient funds to buy them."[813] Peter A. Bod, who would go on to head the Hungarian National Bank, argued in 1990 that privatization would have to include a preliminary phase for extensive building of market institutions.[814] Some Westerners agreed. One Merrill Lynch executive in Budapest suggested in early 1990 that moving slowly would ultimately result in a greater return on Hungarian assets being sold. It would allow time for institutions and markets to develop organically and minimize the risk of failed deals.[815] Other Westerners, pointing to the political sensitivities of some nationalists, suggested that although speed was important, "maintenance of the balance between speed and fairness is critical, because the political risks of privatisation are great."[816]

Many advocates of measured steps toward privatization argued that moving slowly to ensure good results would also solidify support for the process. As Deputy Prime Minister Laszlo Antall suggested before the 1990 elections, "Success would demonstrate to the people they can be winners."[817]

In the end, the MDF won the first round of the debate and with it the election, but privatization moved much slower than even its goals called for.[818] What was the result? In 1990, some were predicting that if privatization did not move quickly, economic stagnation would result.[819] Indeed, if one believes the official statistics,[820] stagnation did occur. The GDP in Hungary declined approximately 8 percent in 1991 and another 2.5 to 3 percent in 1992. Industrial output also slid 10 percent in 1990,[821] more than 14 percent in 1991,[822] and an astounding 15.6 percent in the first seven months of 1992.[823] But

The Ongoing Debates 137

it was unclear whether the pace of privatization was responsible and whether speedy privatization would have yielded other, even worse, political or economic effects. As a result, the debate intensified through 1992.

At times the government policy seemed to reflect deep divisions in its own ranks and ambivalence on the part of policy makers about the proper pace privatization should take. The disappointing slowness of privatization (even according to MDF goals) from 1990 to 1992 was almost certainly attributable to overreaction by the MDF to early scandals. The formation in 1990 of the SPA was in response to the frenetic spontaneous privatizations under the HSWP regime. Then, after the IBUSZ flotation scandal of 1990 discussed above,[824] the head of the SPA, Istvan Tompe, was removed, and his successor Lajos Csepi was reported in July 1990 to want to slow the pace of privatization.[825] The opposition continued to complain, especially when it was announced that the auctioning of small retail shops and restaurants would be closed to foreign bidders.[826] Indeed, the bureaucratic crawl at the SPA was a subject of criticism both inside and outside the government.[827] By June 1991, the government's policy called for accelerating privatization and increasing foreign ownership of Hungarian assets.[828] This new lack of concern by the MDF for foreign ownership of Hungarian assets was a direct result of the disappointing pace of privatization.[829] It also shows that the subsequent attempts to increase domestic purchases of privatization shares were motivated not by any principled objection to foreign ownership but by the simple need to increase the number of buyers.

By January 1991, even Lajos Csepi, who had 18 months earlier wanted to slow the pace of privatization, was telling reporters that the SPA would have to put speed first.[830] This turnaround explains the midcourse correction of 1991 leading to more flexible privatization programs[831] and also explains the shift to boost domestic demand.[832] The MDF acceleration of privatization was due less to a shift in philosophy than to pragmatism: state-owned companies were rotting on the vine while the SPA dithered and held out for maximum price,[833] a fact lamented by one Western accountant:

> The delay has meant these state companies have carried on bleeding to death.... I have seen so many companies, which if they had been able to be privatized 18 months ago, might have sold. Now we come to them and the valuation of these companies has dropped significantly.[834]

Attempts by the SPA to accelerate privatization did not end the debate over pace. The leading opposition groups continued to press for structural changes in the process that would move assets even faster, changes such as putting an end to competitive bidding for companies and restricting the role of the state in the process. Others continued to push for giving more property away as a method of speeding up privatization. Economist Gyorgy Szakolszay told a joint meeting of Germany's Konrad Adenauer Foundation and Hungary's Privatization Research Institute in October 1992 that it would take 5000 years to complete the privatization process at the current pace. He advocated the giving of shares to public institutions, such as universities and the social security system, as a method of creating a private economy. The debate over the pace of privatization would clearly continue.

Measuring and Assessing the Slow Pace of Privatization

The dominant perception in Hungary in the early 1990s was that privatization efforts were only creeping along. The SPA was nicknamed the "State Procrastination Agency."[835] And a joke was circulating in Budapest that before reforms there was light at the end of the tunnel; now there is only tunnel at the end of the light.[836] How slow was the pace of Hungarian privatization in the early 1990s? And what costs and benefits resulted from that pace? These are the questions addressed by this section. As we see, the Hungarian experience offers cautions against overoptimism of policymakers engaged in planning privatization programs. It also points up some important dangers of overcaution and a couple on potential risks of undue haste.

Projections not met. One theme of the story of Hungarian privatization is the inability of the government to meet its own expectations concerning results. We saw this in examining the First and Second Privatization Programs. In addition, revenues from privatization were unimpressive each of the first three years of the 1990s. For 1990 the government projected privatization revenues of $73 million;[837] it raised only $8 million.[838] For 1991 the government first budgeted $2.6 billion in revenues from privatization[839] and later reduced that projection to $540 million.[840] It raised only $418 million.[841] The only year revenues exceeded projections was 1992.

Revenues amounted to approximately $780 million[842] against projections of just over $500 million.[843] The early projection for 1993 revenue was $626 million,[844] but that was revised in December 1992 to $500 million.[845] If the projection for 1993 were on target, the total proceeds from privatization sales in the first four years of privatization would be only $1.7 billion, or between 4.6 and 6.4 percent of the estimated 1990 value of state-owned enterprises. But those estimates were based on book value, which proved to be much greater than actual market value.[846] In fact, the market value of preprivatization state assets was estimated in late 1992 to have been only approximately $6.7 billion.[847] That would mean that if the 1993 projections held true, just over 25 percent of state assets would have been sold in the first four years of privatization. Unfortunately for the SPA, most of the most attractive assets had already been sold, and the revenue trend was downward.

The first head of the SPA, Istvan Tompe, had wanted 80 percent of state-owned assets privatized in just a few years.[848] That clearly did not happen. If the government was going to meet its oft-stated goal of reducing the state sector's input in the economy to well below 50 percent by the mid-1990s,[849] it could not realistically expect privatization to play more than a minor role. Entrepreneurial activity (both domestic and foreign) and the rapidly declining values of the state assets remaining in state hands[850] would together be the prime movers toward dominance of the private sphere as a percentage of total Hungarian economic activity.

These disappointments in privatization were indicative of unrealistic expectations on the part of the government. They also demonstrated a failure of method: projections could have been met had the MDF been willing to commit itself first to pace and only secondarily to political principles. One wonders what interests were served by moving so slowly. The only obvious benefit was to the MDF coalition, which could simultaneously claim to be in favor of privatizing the economy and claim to be preventing the sale of valuable Hungarian assets too cheaply to foreigners. The benefits to the Hungarian economy in general and to the state's budget in particular were less obvious.

The price of slowness. There clearly was a price paid by Hungary in moving so slowly. Delays both chased off would-be purchasers and resulted in the substantial diminution in value of many state-owned companies. This should have surprised no one. As early as April 1990, Sandor Dekany, Hungarian manager of the medium-sized electronics firm Vitali, which had recently spontaneously privatized, issued a warning: "If we take it slowly, many of these companies will collapse and we'll lose out on the opportunities that are now available."[851] Indeed, companies did collapse as Dekany predicted. The examples of the companies Caola, Vegytek, and Danubius Hotels have already been described.[852] In each case, delays in privatization resulted in multimillion dollar declines in value.

Other examples are numerous. The delays in privatizing Danubius Hotels were symptomatic of problems in hotel privatizations generally. In March 1990, the three state-owned chains (Danubius, Pannonia, and HungarHotels) were together projected to be sold at between $693 million and $733 million.[853] All three were placed among the twenty companies in the much-ballyhooed but ill-fated First Privatization Program.[854] In November 1991, delays in privatization had caused that figure to be revised downward to between $467 million and $533 million.[855]

Delays in the sale of Videoton, a large consumer appliance manufacturer pushed that firm to the brink of bankruptcy. The original offer for the firm was $175 million, but there was such outrage expressed at the low price that the deal was rejected.[856] No other bid was forthcoming until late 1991 when the firm's assets were worth between $62.5 and $87.5 million and its debts totaled $375 million.[857] In a frantic bid to avoid bankruptcy, the SPA sold the firm's assets for $50 million, with a 70 percent stake going to the Hungarian Credit Bank,[858] in which the state owned a controlling 49.3 percent block.[859] Not only did the SPA's delay result in a 71 percent decline in the value of the firm, but in the end a state agency (the SPA) sold a majority of the firm to a bank controlled by the state. This is hardly paradigmatic privatization.

According to Hungarian government estimates, by the end of 1992 unprivatized state assets were losing approximately $72.3 million in value every month.[860] The ironic result of the government's initial privatization policy was that by going slowly in an attempt to achieve maximum prices through competitive bidding and to avoid criticism that state assets were being sold to rapacious foreigners at too low a price, Hungary lost untold millions of dollars of potential hard currency revenues.

The price of speed. But of course the SPA is correct that in some instances haste does make waste, as past experience has shown. In fact, one of the problems with the pre-SPA spontaneous privatizations discussed earlier was that assets were being privatized too quickly before anyone was able to assess a value at all. A lot of acrimony resulted when the German press giant Axel Springer traveled around the Hungarian countryside buying up the provincial dailies for seemingly low prices.[861] Critics also point to the IBUSZ public offering discussed above,[862] in which the public offering was vastly oversubscribed, leading some to believe that the deal had been put together too hastily without sufficient assessment of value by the investment bankers involved.

Critics also cite the HungarHotels transaction of late 1989. In December of that year, HungarHotels converted to a joint stock company and sold a 52 percent stake in itself to Quintus, a Swedish/Dutch consortium, for approximately $100 million. A scandal ensued concerning what appeared to many to be too low a price. Ultimately the sale was successfully challenged in court on the grounds of gross undervaluation and failure to consider a competing bid. This delay in the privatization of the hotel chain proved fortunate in the long run, when two of HungarHotel's many properties were sold off for more than $200 million.[863]

Although in some individual cases going as slowly as the SPA went did prove worthwhile, in most cases nothing was gained. And in many cases much was lost. Thus, by the end of 1992, the policy of the SPA was clearly and publicly to accelerate the disappointingly slow pace of privatization.[864]

Policy at the SPA contributing to pace problems. Once it became clear that the pace of privatization was not at an acceptable level, the question became what to do about it. As discussed above,[865] one response of the SPA was to seek to boost demand for state assets by offering credit and other incentives to individual investors, particularly domestic individuals and by moving to privatization mechanisms other than straight sales of companies as going concerns. But the pace problem was only partially a result of lack of demand. As discussed below,[866] there were deep political currents rooted in a fear of foreign exploitation of the Hungarian economy that were dragging down the pace of privatization. There were also problems at the SPA that contributed to the slow movement of state assets.

One problem was simply the slow turnaround of work at the SPA. Many considered it too sluggish.[867] The SPA was always badly understaffed, and what staff it had often lacked the necessary language skills and business expertise to function properly. The corollary was that the SPA had too much work to do. As a result, in the summer and fall of 1991, there was talk, born of frustration, about radically changing the privatization process and reducing, eliminating, or restructuring the role of the SPA.[868] Much of the criticism came from the Ministry of Finance, which seemed to covet the SPA's responsibilities and resent the autonomy and importance of the latter agency.

Finally, in October 1991, the SPA and Ministry of Finance arranged a compromise whereby the asset management functions of the SPA would be transferred to a separate entity (initially called the State Privatization Institute, later transformed into the State Asset Holding Co. PLC).[869] The SPA would thus be free to concentrate on the process of privatization without worrying about managing the portfolio of state assets that would remain in state hands.

The biggest problem at the SPA, however, was a policy problem: it insisted on selling assets only by competitive tender. Despite its liberal rhetoric, which emphasized the SPA's willingness to accept any good offer that was made, the SPA put bids though a complex and lengthy course of evaluation that included the seeking of other offers.[870] Any bid that came in (even in self-privatization and investor-initiated privatization) could be accepted only after competing bids were solicited. Even then, the highest bid could be rejected by the SPA, which, sensitive to right-wing political criticism, would focus on irrelevant asset valuations that often had no relation to market pricing.[871] This policy of insisting on competitive bids without any assurance that the highest bid would be accepted, referred to by one consulting firm executive as a "failure of method,"[872] chased away many would-be purchasers. They refused to sink costs into identifying opportunities only to end up having played stalking horse for another bidder who could ride free on the first bidder's sunk costs in identifying the opportunity. It also slowed the process down miserably. When critics of the slow pace of privatization are asked about problems with the SPA, this is the policy they most often cite. As one accountant explained: "The SPA is swamped. For everything to go through the competitive bidding process is silly."[873]

What, though, could replace the policy? In a rapidly changing economic setting where past performance means very little, how can one assess a bid except through an auction process? One possibility would be for the SPA to ask the bidder for the basis of its valuation of the firm. That valuation, together with what the SPA knows about the firm, could provide the basis for bilateral negotiations. The SPA could also agree to a "no shop" clause, agreeing not to seek competing bids during the period of the negotiations, perhaps for a fee paid by the bidder.

If everyone agreed that privatization should move more quickly, why did it move so slowly? In part, this is simply an acknowledgment of the simple political fact that many Hungarians both inside and outside government still distrusted foreigners, who were doing most of the buying in these privatization sales, as discussed below.[874] But the pace seemed even slower than the SPA said would be ideal. In fairness to the SPA, it encountered numerous problems, many of which could not have been foreseen by officials. Those problems are set forth in a separate section below.[875] But one possible explanation for the SPA's moving more slowly than it seemed to want to lies in the nature of the bureaucracy within the SPA.

An individual decision-maker can be criticized sharply if he approves a transaction that later turns out to have been at too low a price. Indeed, he may lose his job, as former SPA chief Istvan Tompe did after the IBUSZ scandal. But one is never singled out in the press for overcaution. Rather, the cost to SPA officials of overcaution (i.e., public criticism) is spread throughout the institution. Thus, although the institution may be criticized for erring in either direction (overcaution or overaggressiveness), individuals are only criticized for overaggressiveness. Stated slightly differently, SPA decision makers had a lot more to fear in selling assets too cheaply than in not selling assets fast enough. That means the incentive structure for individual decision making within the bureaucracy lies in favor of overcaution.

To overcome that bias, one must either reward aggressiveness (perhaps by giving incentives to bureaucrats for deals completed), punish overcautious behavior (difficult to do), or take away the sanctions for overaggressiveness (perhaps by means of a government directive to the SPA that it put speed ahead of price, thus deflecting somewhat the criticism away from the SPA). Until one of these three things is done, or until the governmnent radically changes its policy favoring sales over giveaways, privatization will not accelerate in any meaningful respect.

How Much Foreign Ownership Can Hungary Stand?

Tied inextricably to the question of pace is the question of foreign ownership. Hungarian policymakers in the early 1990s were constantly faced with the following question: Given the lack of credit and cash resources of domestic Hungarians to participate in privatization, was foreign domination of the privatization process a problem?[876] The answer is: it was, but it should not have been. That is, the problem was one more of perception than substance.

The Problem Created by Perceptions

There can be no doubt that there exists in Hungary, both among the populace generally and in high government circles, the perception that foreign domination of the privatization process is problematic. Before the SPA was created, there was widespread fear that moving too quickly on privatization would mean a rapid sell-off to foreigners by the nomenclatura at too low a price.[877] Early in the privatization process, Hungarians worried that Westerners were profiteering at their expense.[878] They were reportedly concerned about a loss of their "economic sovereignty."[879] As an official of the Hungarian National Bank stated in early 1990, "Many people feel that the country is being sold off and that they are being cheated."[880]

Even later, after it became clear to many that privatization was moving too slowly, the government continued to heed suggestions that foreign purchases were a problem in the eyes of many Hungarians. George Hollo, an SPA official speaking in 1992, highlighted the profoundly political need to involve more Hungarians in the process:

> If all we do is sell to foreigners, that creates problems. A lot of Hungarians don't like it. Down the road if all the large corporations are owned by foreigners and all the small ones are owned by Hungarians, that could create problems.[881]

Lajos Csepi, managing director of the SPA, echoed those thoughts in October 1992 in a comment on the fact that by his estimates 70 to 80 percent of the privatization sales were made to foreigners: "On the one hand this is very positive because it reflects the interest of foreign investors toward Hungary, but in the long run it must be changed, because such a high proportion of foreign capital cannot be maintained."[882] And in fact, the public seemed ambivalent about selling off state assets in a process so dominated by foreign buyers. A 1992 poll showed that 39 percent of Hungarians favored privatizing state assets, but 34 percent were opposed.[883]

Those perceptions that substantial foreign participation was cause for concern dictated government policy and are probably most responsible for the pace privatization took in the years 1990 through 1992. Much of the economic debate during the election campaign that led to the MDF coalition taking power in the spring of 1990 focused on the pace of privatization. The MDF won a plurality of the vote in part because it took the position that privatization should move slowly and under tight governmental control to prevent abuses and domination by foreign corporations.[884] The two leading opposition parties (SDS and Fidesz) had argued for a more rapid privatization, suggesting that foreign ownership was not only necessary but desirable. Because the MDF won a plurality and formed a coalition with the help of the nationalist Independent Smallholders party, no one should have been surprised that the SPA was set up to get control of the process and head off the "potential for foreign exploitation." As one senior government official said not long after the MDF coalition took power, "Hungarians voted for a free market future, but what they didn't vote for was foreign purchase of that future."[885] There clearly existed on the part of government officials a fear of a public backlash if sales were made too quickly to foreigners.[886] This fear of foreign domination in part explains the inordinately slow pace of privatization, as discussed above.[887]

Reality: Some Data Concerning Foreign Ownership and Justifications for Significant Foreign Participation in the Process of Privatization

But Hungarians need not have feared foreign domination of the process. Indeed, there are many reasons they should have welcomed it.

First of all, when the process started in 1990, Hungary had an extremely low level of foreign ownership in its economy in comparison to its own past and in comparison to other countries similarly situated. At the beginning of 1990, only approximately 1 to 2 percent of the Hungarian economy was owned by foreigners.[888]

After a year and a half of privatization and several years of courting direct foreign investment, that figure was up to 3 to 4 percent in June 1991.[889] Before World War II, that figure in Hungary stood somewhere between 20 and 30 percent.[890] Other small Western European countries show foreign ownership rates in the 30 percent range as well.[891] In Austria, for example, economists estimate that approximately 40 percent of the economy is foreign-owned.[892] In Canada, the figure is even higher.[893] In fact, a committee of the Hungarian government's Economic Cabinet suggested in June 1991 that it would be desirable for Hungary to move to the 25 percent level by 1994.[894]

Second, as has been pointed out by observers both outside and inside the Hungarian government, there are strategic benefits to foreign ownership of a larger percentage of Hungary's economy. The SPA itself published a document in March 1991 explaining that due to limited domestic demand, foreigners could be expected to dominate the privatization process, and it suggested the following advantages for the Hungarian economy:

1. Modernization of facilities and methods
2. Encouragement of technical development
3. Reform of management techniques
4. Financial stabilization (i.e., rescue) of some firms
5. Opening of new markets[895]

In addition, Hungarians would benefit from the research and development, new products, and modern marketing techniques that foreign firms would bring.[896]

But the most compelling reason Hungarians' fear of foreign domination of the privatization process was misplaced was a practical one. That foreign participation was necessary; the privatization process simply was not possible without its being almost completely dominated by foreign buyers. As stockbroker Andras Simor put it: "Large companies cannot be sold to widows and orphans.... There is no option but to sell them into foreign hands."[897] This fact was recognized by Hungarian government officials early in the process. Dr. Janos Martonyi, then the commissioner in charge of privatization under the HSWP government, explained that lack of domestic demand would make foreign involvement essential.[898]

As discussed above,[899] an important reason for the low level of domestic participation in the privatization process was a lack of desire: most Hungarians did not want to invest in shares of privatized companies. There have been, however, notable exeptions in which Hungarians have participated in the privatization program. One of the ten largest privatizations of 1991 went to Hungarian investors.[900] The purchase of the Kanisza brewery by its employees and managers was discussed above.[901]

In 1991, Janos Palotas, president of the Hungarian Association of Entrepreneurs, successfully bid for Pharmatrade, an import/export firm specializing in medicinal herbs and basic materials for the pharmaceutical and cosmetics industry.[902] What raised eyebrows was that he beat out the competition, a German bidder, by bidding 824 million HuF ($10.3 million), winning by only 4 million HuF ($53,000).[903] There are occasional sly smiles when the deal is discussed.

Perhaps the most highly publicized example of Hungarian participation in the privatization process was the purchase in 1991 of a majority stake in the Dorog Coal Mines PLC by a group of Hungarian investors. Hungary had not had a privately owned coal mine in 45 years. But in June 1990, the Dorog Coal Mine, then still state-owned, went bankrupt due to poor management.[904] Its 2,300 workers were faced with redundancy. In 1991, however, the coal mine was converted to a company limited by shares, and a 54 percent stake was purchased by a group of Hungarian investors. The group promised to lay off no more than 10 percent of the work force.[905]

But even if they had wanted to, Hungarians could not have participated at significantly greater levels than they did. In 1990, the market value of all Hungarian state-owned assets appears now to have been approximately $6.7 billion.[906] But net private savings in Hungary amounted only to approximately $10.8 billion in mid-1992, including securities, forints, and hard currency.[907] Interest rates on bank deposits was still in the 20 percent range, and investment funds were still virtually unknown. Thus it would have been unrealistic to expect Hungarians to participate at a level higher than the 20 or 30 percent level they did. To rise above that level, Hungarians would have had to commit a higher level of savings to shares or substantial credit resources would have been necessary. But those resources could not have been forthcoming from the Hungarian commercial banking system, which was itself state owned, in need of privatization, and hopelessly backward. Only the state could have provided the credit, and the terms would have had to be subsidized. Indeed, when the government did decide to boost domestic demand in the privatization program, it offered the shares of privatized companies on subsidized terms. And as has already been pointed out, to the extent that an asset's sale is subsidized, that asset is given away. Thus, without a developed commercial source of credit and without sufficient domestic savings, the government was placed in the position of either keeping the state-owned assets and not privatizing them, selling them mostly to foreigners, or giving them away outright or through subsidized sales. At the outset the government chose the second option, then in 1992 it started drifting toward the third as foreign demand began to taper off. Selling primarily to domestic investors was never an option.

Should Shares Be Sold or Given Away?

As discussed above, when the privatization program was still in its infancy, there were calls from many quarters to give state property to the people rather than try to sell it.[908] Hungarian policymakers were thoroughly familiar with the Czech and Polish privatization programs and steadfastly maintained that they would not be engaged in the free distribution of state property except pursuant to limited compensation legislation.[909] Even when the government began subsidizing domestic purchases of privatized property, it continued to maintain the dubious distinction between such subsidized sales and giveaways.[910]

Firmer conclusions regarding which is the better route (sales or giveaways) will have to wait until the Hungarian, Czech, and Polish privatization programs are complete and can be compared. Nonetheless, the three years of privatization in Hungary (1990-1992) do make possible some preliminary conclusions:

1. A realistic valuation of state-owned companies in a country whose economy has suffered under command Socialism for decades ought to eliminate any hope that the state can quickly reap enormous gains by selling those companies. Hungary had hoped to pay down substantially (if not pay off) its massive foreign debt through the sale of state-owned enterprises. But these hopes were based on book values for the state-owned companies in the $30 billion range.

As shown above, market values turned out to be mere fractions of that figure.[911] Former Socialist countries must realistically value their assets and not expect to reap a financial windfall by deciding to sell rather than give away state assets.

2. The economy is not well served when productive assets are given to people who do not have sufficient expertise to manage them efficiently. The overriding problem of the Hungarian economy is that its assets have been poorly managed by the state for approximately 40 years. The economy will not improve if the assets are poorly managed by private persons. Any giveaways therefore should be to persons with sufficient business experience to manage those assets efficiently.

3. Assets would probably be managed better by persons who have some financial stake in the enterprises than by persons who receive state assets wholly gratis. This suggests that if a giveaway program is adopted, the state should consider subsidized sales (i.e., partial giveaways) rather than outright gifts.

4. If sales are deemed desirable, that will almost certainly entail foreign domination of the process, since domestic participation in former Socialist countries will be limited by the low levels of domestic savings in those countries. This would not be the case, however, if the country had a well-developed commercial credit system that could loan domestic investors money at reasonable rates of interest.

5. Gifts of assets can be made relatively quickly, whereas sales take more time. Sales in which the state holds out for the highest possible price take longer than sales in which the first reasonable price is accepted. In fact, holding out for the highest possible price may actually diminish the price received in the end if the state-owned enterprises are deteriorating, as was the case in Hungary. A government embarking upon a privatization program must decide on a timetable for the establishment of a primarily private enterprise economy. It may have to sacrifice price for speed to stay on that timetable and develop the benefits that result from having a private economy. Those benefits are considerable, and the establishment of a private economy quickly may yield greater benefits than the insistence upon sales at a high price.

6. Finally, the issues of pace, foreign ownership, and sales vs. giveaways are inextricably intertwined and must be resolved together.

Should Privatization Be Through Public Offerings or Private Placements?

Because Hungary opened a securities exchange at approximately the same time it began its state-sponsored privatization program,[912] policymakers there were faced with the question whether to privatize companies by floating their shares on the exchange or to privatize companies by negotiating private placements of those firms.

On one level, the question was whether to favor institutional ownership or individual ownership. Hungary recognized the need to create a broad middle class to stabilize the country's reforms politically, and dispersal of share ownership would be one way to move toward that goal. On the other hand, as one stockbroker commented, "For the majority of companies we need active shareholders."[913] Desperately needed managerial expertise and market discipline would require institutional shareholders. As a practical matter, Hungary, like any other country, needed both individual and institutional ownership. Perhaps by accident, that was what it got.

The apparent accidental nature of the dual nature of the Hungarian privatizations is indicated by the confusion regarding whether public offerings were desirable and by the lack of any consistent pattern to the split between privatizations by public offer and privatizations by private placement. According to then industry minister Peter Akos Bod, speaking in June 1990, public offerings on the Budapest Stock Exchange were not to be the main vehicle for initial privatizations because of the small size of the fledgling exchange and because not enough companies were worthy of exchange listing.[914] Yet according to an SPA director, 14 of the 20 companies in the First Privatization Program were initially slated for privatization through the exchange.[915] In later press reports, the SPA confirmed that its initial plan was to privatize through public offerings on the BSE.[916]

Plans to privatize on the BSE did not work out, however. By April 1991, only three companies had been privatized on the BSE,[917] and one of those was the scandalous IBUSZ public offering.[918] By August 1991, only 6 of the original 20 First Privatization Program companies were still thought to be exchange candidates, and none of the 20 had yet been privatized.[919] In September of that year, SPA Managing Director Lajos Csepi said that the SPA was committed to the public offering of privatized companies but that using the BSE was made difficult by the exchange's limited ability to absorb the volume.[920] The Budapest Stock Exchange was being bypassed as a vehicle for privatizations,[921] and policymakers acknowledged that expectations were not being met.[922] There was some talk of privatizations being offered to the public over-the-counter rather than through the BSE.[923] But by late 1991, public offers of any type were nonexistent; all transactions were being worked out on a private placement basis.[924]

The biggest disappointment came when, in November 1991, the planned public flotation of Danubius Hotel stock on the Budapest Stock Exchange was scrapped.[925] The reasons for the collapse of that transaction are explored above.[926] Pulling the transaction was not only a final blow to the First Privatization Program, of which Danubius had been an important part, but it was a blow to the BSE well.[927] The exchange was not growing as expected, and many at the exchange were hopeful that the Danubius transaction would provide a much needed lift.[928]

In fact, of all the big privatizations of 1991, only one — that of clothing manufacturer Styl — ended up with shares on the BSE.[929] And the Styl transaction was really a private placement. It was converted to a joint stock company in 1989, and later the SPA negotiated a private placement of a majority of its shares to GFT Baumler, AG of Germany. The SPA retained 13 percent of the shares, and only after GFT Baumler placed some shares on the BSE did the SPA sell its shares there.[930] Similarly, the ceramics firm Zalakeramia ended up on the Budapest Stock Exchange after a private placement of its shares. A British firm had purchased a 52 percent stake in Zalakeramia, and the firm's shares were listed on the BSE. Only then did the SPA offer its minority interest on the exchange.[931]

As described above,[932] there were two highly successful public offering privatizations floated on the Budapest Stock Exchange in late 1992 — Danubius Hotels and Pick Szeged, the salami manufacturer. Officials issued optimistic statements about the revival of that method of privatization. One even suggested that as many as 30 public offerings could be made on the BSE in 1993.[933] The ultimate mix of public offer/private placement privatizations was impossible to predict. Hungary did in fact develop a mix in the period 1990 through 1992, but it clearly did so without a conscious plan.

Was the Hungarian mix an optimal one? That is difficult to

judge. Dispersal of ownership, participation of small investors, and development of the local securities exchange all militate in favor of public offerings rather than private placements. But not all companies being privatized will be proper candidates for a securities exchange listing, and the capacity of the exchange to absorb the new shares may limit the usefulness of that mechanism in many instances. In addition, private placements take less time and may be much more attractive to investors who believe there is a need for majority control of the privatized firm. In other words, as Hungary discovered, sometimes private placements are better, and sometimes public offerings are better. That decision must be made on a case-by-case basis.

Should the Process of Privatization Be Centralized or Decentralized?

The Hungarian experience offers lessons that many other attempts to privatize large amounts of assets do not because Hungary went through periods of nearly total decentralization of the process, periods of tight governmental control, and a middle ground all in the course of three years. Thus, when asking the question how tightly and centrally controlled the process of privatization should be, the Hungarian experience should be uniquely valuable.

When privatization began under the auspices of the last HSWP government in 1988, the government had effectively "privatized the privatization process."[934] It consciously adopted a policy of noninterference as the managers of state-owned companies privatized them.[935] As described earlier in reviewing the pre-SPA spontaneous privatizations,[936] this completely decentralized process was problematic both economically and politically. The state was not receiving the value it deserved when the companies were sold, and the salesmen were the Communist party nomenclatura, who were lining their pockets as they receded from political power. After the many scandals of that period, the MDF coalition came to power in the spring of 1990 having promised parliamentary control of a centrally managed privatization process.[937]

By late 1990, with the First Privatization Program floundering and the privatization process as a whole far behind schedule, observers noted that the process needed momentum.[938] The process was being too tightly controlled by a rigid SPA policy that was insisting on unrealistic prices. One government economic adviser argued then that the process should be sped up through decentralization of decision making and a return to self-privatizations.[939] But the government was hamstrung by its rhetoric and policy as announced in the election campaign less than a year earlier.

By the summer of 1991, however, the lack of privatization success was becoming somewhat of an embarrassment for the government, and subtle changes in direction began. In early June, the Ownership and Privatization Committee of the government's Economic Cabinet published a draft policy that called for decentralization of the privatization process and accelerated privatization for small and medium-sized firms. At the same time, however, the draft policy reiterated the government's commitment to slow, steady sales at "market" prices.[940]

The Finance Ministry reacted negatively to the draft policy, so it issued its own version approximately two weeks later. The Finance Ministry proposal called for a reduction of input by the SPA and a decentralization of the process by permitting state companies to privatize themselves with the assistance of state-approved advisers.[941]

The Finance Ministry won that battle, and its proposal was the only one considered by the Economic Cabinet.[942] As a result, the SPA announced in July 1991 the first phase of its Self-Privatization Program, whereby certain companies could privatize themselves by selecting an adviser from a government-approved list. The SPA's role would be limited to approving or disapproving any deals actually entered into by the company and a purchaser.[943]

This more decentralized process was widely acclaimed and was said by government officials to be the preferred method for future privatizations. The process still moved disappointingly slowly, but only partly because of the continued centralized control of many transactions at the SPA. There were other problems encountered by the SPA as well, which are catalogued below.[944]

One reason the process continued to be controlled largely at the SPA was government pressure on the SPA to keep control there. In January 1992, the SPA took the public position that its control of the process was justified only in certain circumstances and that in many others it was just a waste of time for the SPA to get too involved.[945] But the MDF government of Jozsef Antall was not so liberal. Beset by deep recession, rising unemployment, and industrial unrest, his government was becoming more authoritarian.[946] The right wing of the party seemed to be getting stronger and more vocal. As a result, the government insisted on tight SPA control in most cases.

Gerd Schwartz was correct when he wrote in 1991 that any country faced with privatizing its state sector needs a strong and independent SPA equivalent.[947] Privatization without state control was a disaster in Hungary, and if it had been allowed to continue, public support for privatization (and perhaps for economic and political reform generally) would probably have eroded. But the SPA in Hungary was not independent enough from the government, and when it became clear to SPA policymakers that the process was overly centralized, they were hindered in their movements toward decentralization by a conservative government policy. More independence at the agency level would have benefited Hungary.

In addition, the Self-Privatization Program should have been expanded. So long as there exist financial incentives for the advisers involved to bring in the highest possible proposal, in most transactions there is no need for the SPA to be involved except as a shareholder exercizing its right to approve or disapprove a transaction negotiated by management. One problem encountered by the SPA was reluctance on the part of some managers to undertake self-privatization. This could have been dealt with more forcefully either by replacing recalcitrant management or by insisting under force of law that the managers take the required steps toward privatization.

Self-privatization is no panacea because as Hungary discovered, there are not necessarily buyers for every firm being sold. Liquidation, piecemeal sales of assets, leasing programs and the like all require more centralized input. But Hungary did not go as far toward decentralization as it might have, and its privatization efforts suffered as a result.

How Should Shares Be Distributed Among the Many Claimants?

Because of the unclear ownership structures of many Hungarian state-owned enterprises,[948] Hungary faced a question that not every country undergoing privatization will face: who should receive shares in a state enterprise being transformed into a share company as the first step toward privatization? In Hungary there were many claimants to "ownership" status. In addition, Hungarian policy called for some participation by employees, and some distribution of property pursuant to the Compensation Law. The queue of claimants was long in many cases. An example from 1992 illustrates the complexity of the negotiations typical of Hungarian privatizations.

When Pick Szeged, the salami producer, was privatized in late 1992 by public offering on the Budapest Stock Exchange, only approximately half the shares were available for sale to the public after the various claimants took their shares. In July 1992, the SPA was proposing the following breakdown:[949]

Local Councils	5.0 percent
Employees	10.0 percent
Compensation Coupons	20.0 percent
Public Offering	75.0 percent

Over the course of the summer and fall, however, the negotiations resulted in a much different breakdown:[950]

State Property Agency	5.0 percent + one share
Employees	10.8 percent
Local Councils	4.1 percent
Compensation Coupons	20.0 percent
Creditors (debt/equity swap)	11.0 percent
Public Offering	remaining shares

The Pick Szeged example is typical. At about the same time Pick Szeged was being privatized, the Danubuis Hotel chain was also being sold by public offering. And as in the case of Pick Szeged, a certain percentage was reserved for the SPA, a certain percentage was allocated to the local councils who had contributed the land for the hotels, a certain percentage was reserved for the employees, and a certain percentage was given to financial institutions to retire debt obligations.[951]

It would be impossible to second-guess the government's allocation of shares in any individual case without knowing exactly the relative postures of each party. The list of claimants and the strength of each claim varies from case to case. Some allocations were made in Hungary as a result of legal claims of "ownership" (e.g., to the local councils), but others were made for policy reasons (e.g., to holders of compensation coupons). What Hungary's experience tells us is that before embarking on a process of privatization, planners not in a position to dictate terms must consider in advance all the possible claimants to shares and be prepared for complex negotiations in individual cases.

How Should the Proceeds of Privatization Be Spent?

The flip side to the question of who should receive shares when an enterprise is transformed into a share company is the question of how the state should spend the proceeds derived from the sale of its share. As with so many things in the Hungarian privatization experience, we see an evolution of policy as Hungary experimented. Early hopes were subdued by the political reality of budget deficits.

The initial plan in Hungary was to use the proceeds from the sale of state assets to pay down (and perhaps pay off) the national external debt, which was nearly $20 billion in the opening years of the 1990s. Indeed in March 1991, the SPA published a brochure claiming that the proceeds from privatization would be used primarily to pay down the national debt.[952] By that summer, however, the government was faced with a deteriorating economy and began to backtrack. A government committee suggested in June that 75 percent of privatization proceeds go to paying off the national debt *and to supporting private ventures, developing the country's infrastructure, privatization support, and job creation schemes.*[953] The national debt was going to have to share.

In August of that year, the government announced that it would plow some unspecified percentage of its privatization proceeds back into the privatized companies to fight unemployment.[954] Later that figure was said to be as high as 20 percent.[955] In September 1991, the external debt stood at approximately $18.4 billion, and only 29 percent of the total privatization revenues received to that date had been used to pay it down.[956] At about the same time, the government was negotiating with the International Monetary Fund for financial assistance, and the IMF was insisting on signs of economic austerity that would bring down the budget deficit. The government attempted to use privatization revenues to show that the budget deficit was under control, but the IMF demurred, pointing out that

Hungary was simply using the privatization revenues to support budgetary profligacy rather than actually tightening its belt.[957]

By February 1992, a dispute arose within the government over how privatization revenues should be spent. The Minister of Finance wanted to pay down the national debt, but the head of the SPA wanted to stimulate the economy.[958] The SPA won the argument: in the first five months of 1992, only 13.1 percent of the privatization revenues was used to pay down the national debt. More than 60 percent was simply credited to the state budget. Here is how those figures broke down:[959]

State Budget	60.2 percent
National Debt	13.1 percent
Local Governments	2.7 percent
Costs of Privatization and Social Security	24.0 percent

Hungary's thinking concerning how privatization revenues should be spent had clearly evolved since 1990.

Did Hungary spend its money wisely? That is hard to say. On the one hand, there is some logic to paying off a mortgage from the sale of the mortgaged property. And Hungary's long-term economic health will require facing that huge foreign debt at some point. On the other hand, privatization and other economic reforms require popular support, and popular support could wane if the effects of reform were not softened to the extent possible. And the huge budget deficit in Hungary was threatening to undermine efforts on the part of the government to bring inflation under control. Still, many remained suspicious that the government was being short-sighted, engaging in government spending just to soften public criticism and keep itself in power.[960]

How Much Residual State Ownership Is Desirable?

A final question that must be confronted by a country going through the process of privatization is how much in the way of productive assets to keep in the hands of the government. Like other questions, the answer to this one will vary from country to country, as vital strategic interests differ. But Hungary clearly needed to keep less than it planned to.

And the number of companies the state planned to keep in state hands expanded over time. In May 1991, the government announced that 19 key enterprises would remain wholly in state hands. The list was concentrated in the "strategic" fields of energy, telecommunications, postal services, air travel, television, aluminum production, railways, and savings banks.[961] The state would also retain a majority stake in another 15 companies, and the agriculture minister listed certain agricultural properties to be retained in whole or in part by the state.[962]

By January 1992, however, the government was saying it would retain as many as 150 enterprises,[963] representing 20 or 21 percent of the economy.[964] Notwithstanding the government's continued assertions that only strategically important firms would remain in state hands[965] and its recognition that "if anybody has learned that the state does not work well as an owner, we have learned it,"[966] the list of "strategically important" enterprises had expanded.

On June 22, 1992, the Parliament passed a law creating the State Asset Holding Co.[967] and the same month published a new expanded list of companies to be retained by the state.[968] The number of retained enterprises was now thought to be as high as 170.[969] The SAHC would be a for-profit company that would manage a portfolio of minority interests the SPA had retained when it privatized firms and the majority interests of firms that would remain in state control.[970] The SPA planned to turn over to the SAHC approximately 78 percent of the minority interests it had retained; the remaining shares would be given to the social security system or made available for compensation coupons.[971] Officials promised that assets held by the SAHC would be available for lease or privatization on a case-by-case basis and that the list of companies to be retained would be reviewed every two years.[972]

Still, not everyone had confidence in the ability of the government to do a better job of managing assets than it had before. Government minister Karoly Szabo resigned his SPA post in January 1992 in part as a result of the government's plans to retain so much productive property. "I don't think the state is capable of being a good manager in Hungary," he told an interviewer.[973] And when the government first proposed the creation of the SAHC in March 1992, the opposition in Parliament criticized the move as just another attempt to concentrate power in the government's hands through "renationalization."[974]

In reviewing the Hungarian example, one must differentiate between those instances in which the state retained a small minority stake in a company as a good investment (e.g., Pick Szeged) and those in which the state retained complete or majority control for strategic reasons (e.g., Malev Airlines). The former are clearly justified on economic grounds if the firms are in fact good investments. The latter are much more difficult to justify. Certainly control over strategic industries (e.g., weapons manufacture, energy production, etc.) is vital. But giving up ownership does not entail giving up control. The power to regulate, which the government always maintains, is the power to control. And Western economies (particularly that of the United States) have shown that a country does not become less secure when vital industries are only regulated rather than owned by the state.

Retaining state ownership to ensure the continued employment of the work force is simply bad policy, ossifying as it does the inefficient use of assets as a jobs program. In the aggregate, the work force fares better if the economy grows, thus generating revenues to cope with localized dislocations. The example of the Borsodtavho firm illustrates the shortsightedness of undue retention of state interests.

In 1992, Borsodtavho was Hungary's second largest heating company, providing residential and commercial heating services throughout a geographic district. Until October 1991, the company received an annual subsidy from the central government, which owned the firm, of approximately $14.5 million. That month the central government decided to unload its interest in Borsodtavho. But instead of privatizing the company, the government simply transferred ownership to the local government for the district of Borsodtavho's operations and eliminated the subsidy. Heat for the populace was too important to be left to the market. But less than one year later, Borsodtavho declared bankruptcy. The result was that thousands of Hungarians faced the winter of 1992-1993 with the possibility of no heat in their homes. In addition, four large suppliers to Borsodtavho were jeopardized by the inability of Borsodtavho to pay them.

Ten

The Problems of Privatization — "The Process Is Far More Complicated Than We Ever Thought"

As should be apparent from all the foregoing discussion, Hungary encountered numerous problems in its attempt to privatize its moribund state sector. That fact does not reflect as badly on the Hungarian policymakers as it would seem, since they were exploring terrain that no country had ever explored before. Some of the difficulties they experienced could have been foreseen, but many could not have been. Some simply could not have been avoided. The Hungarian experience will at least alert future countries similarly situated to likely problems that should be planned for from the outset.

This section presents a systematic catalogue of the difficulties that hindered Hungarian privatization attempts. Many of these have been discussed extensively above, and where that is the case, they need only be mentioned briefly here. Other problems, not fully analyzed above, are discussed more fully below.

The Political Problem of Foreign Ownership

It is impossible to tell how much a role the discomfort over possible foreign domination of the privatization process, viewed in isolation from other problems, slowed down Hungary's privatization. As we have seen,[976] that discomfort shaped many aspects of the program. It led to centralized control of the process and to a policy that emphasized slow, overcautious review of bids. It also led to complex devices (share set-asides for domestic purchasers, special easy credit programs, separate Hungarians-only auctions of small shops and restaurants, etc.) that stole the time and attention of the privatizers away from their main task: putting state assets into private hands. But the fact is that Hungary did court foreign investors and allowed them to purchase between 70 and 80 percent of the assets being privatized. The political problem of foreign participation was much greater than it should have been, but not as bad as it could have been. Other countries planning a similar program must calculate at the outset how much foreign participation is desirable (and/or politically tenable) and set policies accordingly.

Lack of Demand

One problem that worsened over time for Hungary was that investors did not want to buy all the assets Hungary had for sale, at least not at the prices the Hungarians hoped to get for them. That was true for both foreign and domestic purchasers.

Domestic

Although before World War II, Hungary had a thriving securities exchange, in post-cold war Hungary there was almost no equity culture at all. A comprehensive social safety net, together with restrictions on a person's ability to generate wealth through private enterprise, conspired to encourage consumption over savings. To the extent that individuals invested their money at all, they tended to invest in government debt instruments and hard assets such as owner-occupied residential real estate. Financial profiteering through passive equity investments was ideologically taboo and a foreign concept. For those reasons and others, domestic demand for shares in privatized companies has been understandably low — between 20 and 30 percent of privatization purchases.[977]

A common perception is that in the early 1990s, domestic Hungarians simply could not afford to participate at any significant level in the privatization process.[978] One commentator has suggested that domestic savings would permit the purchase of, at most, 10 percent of state assets, even if credit were available.[979] Or as one stockbroker explained the need to sell to foreigners, "Large companies cannot be sold to widows and orphans."[980]

But this perception that Hungarians could not afford to participate is, strictly speaking, incorrect. Theoretically, Hungarians had (by 1992 anyway) the savings to participate at a higher level than they did.[981] Statistics published by the Hungarian National Bank show that Hungarians had in August 1992 net savings (i.e., bank savings minus personal loans) of approximately $10.8 billion.[982] Given that state assets were worth approximately $6.7 billion when the privatization process started[983] and that total privatization revenues for 1993 are expected to be about $500 million,[984] Hungarians could generate more demand for privatized shares than they do. But that would require them to commit, en masse, a very high percentage of their savings to share purchases in former state companies.

Although Hungarians in the aggregate do have enough money to generate more demand in the privatization process, there is a dispersal of wealth problem. The new bourgeoisie—that stratum of Hungarians with liquid risk capital—was described by one banker as "fragile and very thin."[985] As another observer noted, the number of Hungarians who could put out several hundred million forints for a state-owned enterprise could be counted on one hand.[986] But Hungary would benefit more from a wide dispersal of share ownership than from a few purchases of substantial assets by a few domestic investors anyway. Hence the need for credit and investment funds for domestic purchases.

But the level of wealth was not the primary reason for the slack demand on the part of domestic investors. The biggest problem was one of will: Hungarians with money did not want to put that money into privatized state firms for a number of reasons. They were unfamiliar with securities markets and share ownership and as a consequence felt insecure in putting their money into those markets. As several commentators have noted, there was in the early 1990s no "equity culture" in Hungary.[987] Given the recent economic and political turbulence, potential investors were more likely to put their savings into useful hard assets (such as housing) or short-term debt securities of the state, which were perceived to be of little risk.

In addition, those with funds and a willingness to invest in shares did not want to put their money in former Hungarian state companies. They could invest hard currency abroad, and forints could be put directly into start-up enterprises. They had seen firsthand how the state firms had been operated and were therefore perhaps more realistic than foreigners about the prospects of those firms.

Foreign

After an initial period of euphoria, foreigners were duly skeptical of the relative value of privatized Hungarian assets in comparison with other investment opportunities worldwide. Demand had always been patchy, with very little demand in some sectors, such as chemical, petroleum, and engineering.[988] In the other sectors, after the best Hungarian firms were sold,[989] foreign demand was slackening off by 1992.[990] As a Merrill Lynch executive observed, "Foreigners may be willing to take the family silver, but they're going to be far less interested in the base metal."[991] With its economic difficulties, lack of financial infrastructure, and proximity to the Balkan conflicts, Hungary was at a disadvantage in the world competition for capital. Westerners claimed they needed a discount to be attracted to Hungarian investments.[992] One Austrian banker was quoted as saying he valued Hungarian assets by considering what comparable Western assets would be worth and dividing by five.[993]

Lack of Credit

The problems created because the banking system of Hungary was unable to provide credit on reasonable terms to Hungarian investors were explored above.[994] This lack of credit created problems not only for the privatization programs stressing sales of shares in large and midsized state-owned companies, but also in the sale of small shops and restaurants.[995] In 1991 and 1992, the government clearly recognized that domestic demand was being hindered by a lack of easy credit and moved to expand the availability of that credit.[996]

Valuation Difficulties

Nearly all participants in the Hungarian privatization process commenting on the problems discovered by the Hungarians mention the extreme difficulty in placing a value on state-owned assets.[997] The problem, which has already been discussed above,[998] was simply this: The former state-owned companies operated in a command economy, within COMECON, with the Soviet Union as a major trading partner. After Hungary moved to a market economy, after COMECON collapsed, and after the Soviet Union dismantled itself and then slipped into economic chaos, past valuations and performance of Hungarian state firms became irrelevant. In other words, past performance figures were artificial, and there was no way to predict what the new future would bring anyway. Hungary was caught in a Catch 22: for purposes of creating a private market economy, it was trying to assign values that only a private market economy could provide. As one commentator noted, "It's very hard to say what's cheap and what's expensive when there isn't a handy yardstick."[999]

This created several problems for Hungary's privatization program. First, it slowed the process down by making more difficult the SPA's task of valuing state companies before selling them. Second, it added another element of uncertainty in the minds of potential bidders, thus increasing the relative riskiness of Hungarian investment in comparison to other potential ventures. Finally, it created conflict in the negotiations between the SPA and the bidders for those companies because without information that the parties could agree was relevant, widely disparate valuations were common. The SPA responded to uncertainty by erring, partly for political reasons, on the side of high valuations, whereas bidders naturally erred on the side of low valuations.

Overvaluation by the SPA

Although as one Western businessman complained, "prices tend to be pretty arbitrary,"[1000] most observers agree that the SPA was asking far too much for the assets it was trying to sell.[1001] Notwithstanding its repeated insistence that it was seeking to sell assets at the market price, the SPA seemed not to understand that if no buyer is willing to pay what the seller is asking, the seller's asking price is not a "market" price. "The Hungarians have difficulty with these concepts," remarked the Western businessman.[1002] The SPA reacted to uncertainty by erring on the high side.[1003] "The Hungarians want unrealistic prices for their companies," noted a Western financier. "They have to understand that there are other places to put money."[1004]

The SPA was reacting to real and potential criticism from the right and thus ignored its liberal rhetoric claiming it would accept the highest bid.[1005] The aggregate effects of overvaluation by the SPA are difficult to measure, although as we have seen, this problem surely slowed the pace of privatization[1006] and also meant the collapse of deals in individual cases.[1007]

Political Distractions

Political infighting and conflicting government policy may also have slowed the pace of privatization in Hungary.[1008] Examples of that infighting and conflict indicate that no one person or agency was fully in charge of privatization policy. We have already seen a dispute over privatization policy between the Ministry of Finance and a committee of the government's Economic Cabinet.[1009] In 1991 there were reports of a split in the government over who should control privatization policy. Some argued that the government should take charge at the parliamentary level, whereas the Finance Minister was reported to want to take control of privatization in the Finance Ministry.[1010] The Finance Minister also argued with the head of the SPA over how privatization proceeds should be spent.[1011] Finally, in February 1993, Finance Minister Mihaly Kupa resigned, in part over his frustration with governmental privatization policy.[1012] Although the effects cannot be quantified, political squabbling almost surely slowed down privatization.

A second type of political problem erupted in 1992 when prominent MDF member Isvan Csurka published an article in a Budapest newspaper with jingoistic and anti-Semitic overtones. Csurka suggested that a Jewish-led conspiracy of former Communists, Western bankers, and the IMF was threatening Hungary's future. He also claimed that there were "genetic reasons" for Hungary's difficulties and that Hungary should seek "living space" among those nations to whom it had lost land after World War I.[1013]

There were large public rallies in Budapest in opposition to Csurka's ideas and counterrallies by pro-Csurka demonstrators. There were calls both domestically and abroad for Csurka's denunciation by the party head, Prime Minister Jozsef Antall, but in a party congress in late 1992, Antall reached an accommodation with Csurka. Csurka never withdrew his remarks, and Antall never denounced them. Given Hungary's past (it had a semifascist government between the world wars and a Nazi government during World War II), the episode made many foreigners (particularly Jewish businessmen) nervous. Approximately 600,000 Hungarian Jews were exterminated in the late 30s and early 40s, so some observers feared that if the economic situation worsened, Hungarian nationalists would again seek scapegoats. Again, the effect of such an episode on foreign investment is impossible to measure, but it certainly did not improve the climate for that investment or enhance the reputation of Hungary as a stable liberal democracy.

Lack of Clarity of Property Ownership

We have already seen the unclear nature of property ownership within so-called "state enterprises" both generally[1014] and in the farming sector in particular.[1015] There can be no doubt that the confusion over who owned what and how much of state property slowed privatization down. Indeed, this problem, together with the problems of valuation and overvaluation by the SPA, were almost universally acknowledged, both within Hungary and among outside observers, as the primary factors slowing down privatization.[1016] As political economist Laszlo Lengyel noted, "Because of the unclear, indistinct nature of ownership relations, it is hard to say where state property ends and private property begins in Hungary."[1017] The sale of the famous Gellert Hotel, part of the Danubius chain, had to be scrapped due to conflicting ownership claims.[1018] When the Danubius chain was privatized by public offering in late 1992, the Gellert Hotel was not included for the same reason.[1019]

This was a problem not only for the SPA, which had to untangle the conflicting ownership claims and decide which claimants were entitled to enter into the negotiations for a share of the company,[1020] but also for the bidders. In self-privatizations and investor-initiated privatizations, the bidder would deal directly with the firm's management, and sometimes it was unclear who had the authority to strike a deal.[1021]

Insufficient Financial and Legal Infrastructure

Few within the Hungarian government or within the ranks of potential foreign investors probably foresaw the difficulties that would result from the primitive and/or non-Western financial and legal infrastructure in place in Hungary when the privatization program began. Peter Akos Bod noted in early 1990 that privatization would require an extensive market-building phase.[1022] Most observers, however, were surprised at the scope of the problems created. The banking system did not function as a commercial banking system does in the West,[1023] and as a result there were difficulties with the transfer of funds and there was a lack of credit for small businesses.[1024] The lack of Western-style intellectual property protection laws made high tech companies unattractive.[1025] The Hungarian accounting system was in the early years not aligned with Western standards and thus valuations were slowed down tremendously.[1026] One accountant remarked on the difficulty in 1990: "We are hopeful that by the end of 1991 there will be accepted phrases for things like nonperforming loans."[1027] A senior government official was quoted as saying, "Until a few years ago, the only reason most companies kept accounts was to give work to another 250 people."[1028]

The thin capital markets also added to the problems of valuation and liquidity.[1029] Some were critical of Hungary's attempt to privatize its economy without first going though the period of bringing its legal and financial infrastructure up to date:

> Privatisation is being introduced back to front—without an infrastructure to support it. None of these countries [in Central and Eastern Europe] yet has a Western-style accounting, banking, or capital markets system. Nor does any of them have much of an indigenous property-owning class.[1030]

In fairness, Hungary did take some important early steps to provide a sound legal infrastructure for privatization. The Act on Securities, the Act on Companies, and the Act on Foreign Investment, all discussed above,[1031] are significant examples. In addition, the establishment of the Budapest Stock Exchange and State Securities Supervision Board were, if anything, premature. The lacking elements of infrastructure must be seen in perspective.

David Bartlett has argued that these various holes in Hungary's infrastructure were interrelated and together hindered the privatization program. The thin capital markets and lack of financial intermediaries made it difficult to privatize through public offerings of securities, and this hampered attempts to mobilize public support. Therefore, the government had to seek direct foreign investment, which made the privatization program politically delicate. In addition, the thin capital markets made valuation difficult, which, combined with the dominance of foreigners, led to a public suspicion that the assets being sold were being undervalued. These various problems all slowed down privatization, leading to a shortage of immediate budgetary funds to soften the effects of privatization and thereby eroding public support even more and forcing a further slowdown in the process.[1032]

Bartlett is clearly correct in suggesting that the infrastructure problems were interrelated. Those who suggest, however, that privatization should have waited until after these changes in the legal and financial infrastructure were in place are being unrealistic. Some of the infrastructure (capital markets, for example) required a system of private property to operate. Other elements, such as changes in the accounting system, were being carried out simultaneously with the privatizations. Privatization might have been smoother had it been postponed, but it would not have come about at a faster pace.

International Conflict

Hungary began its privatization program at an inauspicious time in world events. Worldwide tensions escalated substantially in the fall of 1990 when Iraq invaded Kuwait and again in January 1991 when an allied coalition began military action to retake Kuwait. Many Western businesses, fearing terrorism, cut back on world travel. Then in 1992 conflicts in the former Yugoslav republics boiled over into armed conflict. Because Hungary shares its southern border with Croatia and Serbia and because a northern section of Serbia (Vojvodina) has a large ethnic Hungarian population, that conflict threatened to spill over into Hungary. These difficulties almost certainly had a negative, but unquantifiable, effect on Hungarian efforts to attract foreign investment. The United Nations boycott of

Serbia also cut off an important export market for some Hungarian firms, thereby making them less attractive to Western investors.

Bureaucracy, Controversy, and Confusion at the SPA

We have already seen how the dynamic of decision-making at the State Property Agency led to overcaution in approving transactions presented to it.[1033] This bureaucratic hesitancy in a new and unfamiliar territory was widely perceived to be one of the problems facing Hungary's privatization program,[1034] and the bureaucracy was criticized both by government officials and foreign investors for being slow, confused, and arbitrary.[1035] Critics pointed not only to the general attitude of overcaution but to specific manifestations of that attitude, such as procedures calling for accountants to determine the value of the firm and rules requiring competing tenders for all companies.[1036]

That these problems existed is undeniable, but it would be unfair to put all the responsibility for them on the SPA. The SPA was sailing in uncharted waters, so for the first couple of years, some confusion would be expected. In addition, the government never placed enough resources at the hands of the SPA to ensure a sufficiently large and qualified staff. The government must also bear the responsibility for the general lack of vigor in selling assets; it was the MDF's stated policy to move slowly and hold out for high prices.

The SPA also cannot be faulted for some of the political distractions that resulted from government displeasure over the shape SPA policies were taking. We have already reviewed the IBUSZ public offering, which resulted in the ouster of Istvan Tompe as managing director of the SPA and his replacement with Lajos Csepi.[1037] We have also noted the resignation of Deputy Managing Director Karoly Szabo over policy differences with the government[1038] and the threatened ouster of Csepi himself in 1992.[1039]

The SPA itself, however, must be faulted for the occasional confusion within the agency. The BZW controversy of late 1992 is a good example. In early November, the SPA took the extraordinary step of scratching the firm of Barclays de Zoete Wedd from the approved list of privatization advisers for alleged wrongdoing.[1040] BZW was the adviser on the privatization of the Kobanya Brewery, which held at the time approximately 33 percent of the domestic beer market. The SPA claimed that by sending prospectuses and other materials to three companies that were not on the approved list of bidders because of their existing significant market shares in the domestic beer market, BZW violated the rules of the closed tender. SPA chief Lajos Csepi referred to it as the most serious breach of privatization rules to date.[1041]

The following week, however, the SPA was backpedaling. After meeting with representatives of BZW, an SPA spokesman acknowledged that the rules had not been violated wittingly.[1042] One source attributed the flap to "SPA incompetence and Western bank arrogance."[1043] But eventually the SPA took full responsibility, noting that its intentions were "not clearly communicated to BZW which led to steps taken by BZW which were found objectionable by the SPA." BZW was restored to the adviser list.[1044] With all the unavoidable problems encountered in the privatization process, the SPA did not need such embarrassments.

The Bad and Worsening Condition of Many State-Owned Enterprises

In order to sell an asset for a high price, one must possess an attractive asset. Hungary had many unattractive assets to sell in the early 1990s, and with the slow pace of privatization, many were becoming less and less attractive as time went by. Indeed, as the privatization program began, there were indications that many in Hungary understood that some firms would be difficult to sell. They spoke of the lack of good prospects[1045] and the fact that some companies were simply too sick to sell.[1046] State companies were described in 1990 as "overmanned, undercapitalized, monopolistic, and terribly managed."[1047] Every firm needed substantial financial, legal, and organizational restructuring.[1048] Overstaffing was a particular problem, with many companies swollen so much in comparison with Western firms that they appeared to be jobs programs rather than profit-oriented companies.[1049]

Many of these problems were essentially managerial and would ordinarily lend themselves to being exploited by a purchase and turnaround by shrewd investors who were certain they could provide better management. But in Hungary two factors inhibited that classic "raider" strategy. The first was the language problem. Hungarian is an obscure enough language that the number of Hungarian speakers with the management expertise to turn a troubled state firm around was limited. Second, many of the state firms were sinking so rapidly[1050] that the shrewd investors were waiting for liquidation, when the assets could be purchased at a lower price without the need to purchase the entire firm.[1051] This may explain why liquidation was in the vanguard of privatization strategies in 1992.[1052] The SPA thus faced a significant difficulty because as soon as a firm appeared to be headed toward eventual bankruptcy, foreign bidders would back off, thereby ensuring the beleaguered firm's demise.

Lack of Qualified Staffing at the SPA

The lack of qualified staffing at the SPA has been explored above.[1053] Providing sufficient resources to hire an adequate staff was the government's responsibility. But there should not have been a need for so many staff members. If the SPA had taken a more aggressive approach to selling assets (i.e., one that focused on selling quickly to the highest bidder and decentralized the workload into the hands of the firms being privatized and their advisers), it would not have needed as many staffers and would have been able to move the more limited workload through even faster.

Possible Conflicts Among Laws

One perhaps unavoidable result of the tremendous volume of new law being spewed forth from the Hungarian Parliamament as it attempted to overhaul completely its legal system was some conflict between the policies represented by those laws. That conflict also highlights the shifting and multifaceted coalition within the government, a coalition whose members were often pushing the government in different directions. Two examples illustrate the problem.

Telecommunications Legislation

On November 23, 1992, after months of debate, the Hungarian Parliament passed a sweeping new telecommunications law.[1054] The goals of the act were to provide a basis for stable development of the telecommunications industry in Hungary and to facilitate the privatization of the Hungarian Telecommunications Company (HTC), which had held the telecommunications monopoly in the country.

The problem is that the terms of the Telecommunications Act raise serious questions whether investors would want to buy shares in the HTC under the terms of the act. Until passage of the act, the HTC held all the local operating franchises to provide telephone service throughout Hungary. The new act, however, raises serious doubts about how many of those local franchises the HTC will continue to hold. Section 4 of the Telecommunications Act requires the Ministry of Transport, Telecommunications, and Water Management to invite competitive bids for any local franchise if such a competitive process is requested by one-half of the local governments in the franchise area. In the case of all other franchises, the ministry may make awards on a noncompetitive basis.[1055]

This competitive bidding process resulted from strenuous lobbying efforts on the part of local governments, who were frustrated by the historical unresponsiveness of the HTC (most homes outside Budapest do not have telephones, and waiting lists are years long). In addition, local entrepreneurs lobbied hard for the opportunity to bid on these franchises, arguing persuasively before the populist/nationalist members of the ruling coalition that local businessmen should not be shut out of the Hungarian telecommunications industry in favor of foreign investors buying HTC shares.[1056]

The result will probably be a significant delay in the privatization of the HTC, which the SPA had hoped to sell by the end of 1993. Investors will be unable to assess the value of the post–Telecommunications Act HTC until they know how many of the local franchises the HTC will own. Once again, the problem of valuation in a rapidly changing legal/business environment will create a problem for Hungarian privatization efforts.

Compensation Legislation

Uncertainty over the effect of another piece of legislation also undoubtedly slowed down privatization efforts in the early 1990s. One of the most important political and economic questions arising out of the reform period in Hungary was how (if at all) the state would compensate former owners for property that had been appropriated by the state. The specific question that caused hesitation in the minds of foreign investors interested in the privatization process was this: Is the property being privatized subject to prior legal claims of former owners? If so, that property obviously would be worth less.

Partly to remove the cloud this question placed over the privatization process, the Hungarian Parliament passed on June 26, 1991, its Compensation Act.[1057] Observers hoped and expected that the act would accelerate privatization.[1058] Indeed, one of the goals of the Compensation Act was to promote business certainty by making clear exactly how former owners could make recoveries.[1059] But the act may actually have slowed down privatization, at least in the short term.[1060] This was because under the terms of the act, real property that had been used as farmland at the time it was taken was to be put up for auction to families whose farms had been taken by the state.[1061]

Thus for privatized companies situated anywhere but in the center of a city, potential bidders either had to wait until the auction process was complete, take a chance that the real property would not be subject to claims of previous owners, or make bids discounted by the possibility of former owners having superior claims. In the long run, the legislation would clearly facilitate privatization, but in the short run, it codified the justification for uncertainty many potential bidders felt.

Environmental Problems

A final problem, not often discussed in connection with Hungarian privatization, was more latent than the others and therefore more insidious. All the Eastern Bloc countries had for years been sacrificing the environment for the sake of industrial production and employment, and in the early 1990s the scope and cost of cleanup was daunting throughout Central and Eastern Europe. This presented two problems to potential investors in privatized companies.[1062] First, they had to make thorough environmental damage assessments because of the high likelihood of environmental pollution in their locations, sources, and products. This represented a substantial entry barrier because the assessment would have to be made before any decision to make a bid could be reached. Second, and perhaps more seriously, the potential bidders had to speculate about what approach would be taken to environmental cleanup in the future. If, for example, the government were to approach the problem by imposing huge cleanup burdens on all operators in Hungary, the effect on future income streams would be substantial. These considerations were by necessity factored into the decision making of potential bidders with the certain result of discouraging bids.[1063]

Eleven

What Do We Make of All This? — Some Lessons to Be Learned from the Hungarian Privatization Process

Although the Hungarian regulatory scheme for privatizing its enterprises makes interesting history in its own right, it also offers important lessons that may be of use to other countries embarking upon substantial privatization programs. This is particularly true of other former Socialist countries in Central and Eastern Europe, which are similarly situated to Hungary. This concluding section attempts to draw on Hungary's experience and list systematically the most important lessons to be drawn:

1. The process of privatization must be centrally controlled by the state to prevent private wealth-seeking by those in control of the means of production and to foster the public confidence in the process necessary to ensure its political viability.

2. To the extent possible, however, the role of the state in the process should be minimized to avoid bureaucratic delays. In other words, although the process should be ultimately controlled by the state, that control should be limited to final decision making, and the day-to-day work of preparing companies for privatization, seeking out and negotiating with possible bidders, and the like should be as decentralized as possible. A country engaging in massive privatization has presumably discovered that government bureaucracies make poor business managers. This is no less true in privatizing

business than in conducting business. Hungary's state-controlled spontaneous privatizations illustrate this point.

3. The state organ in charge of privatization must be strong and independent of fleeting political squabbles and criticism. It should be insulated from political debates to the greatest extent possible to permit it to conduct its work according to the overall government plan without fear of undue criticism over inevitable occasional failures. To that end, the agency's decision makers might be appointed on a nonpartisan or multiparty basis and guaranteed job tenure for a period of years.

4. The government should commit sufficient resources for the agency in charge of privatization to do its work. From the outset, that agency should be sufficiently staffed with persons possessing the requisite business and language skills to work efficiently. Privatization efforts are hindered when a small staff learns on the job.

5. Privatization without political stability would be difficult at best. Hungary successfully showed potential investors that its political reforms were irreversible. Before embarking on a privatization process, other countries would be wise to establish an acceptable level of public commitment to political stability as well. Similarly, there must be strong public support for the process of privatization, even with its negative aspects, or the process will be jeopardized.

6. Certain pieces of legal infrastructure should be put in place before privatization efforts begin. For example, a company law authorizing and regulating private share companies and their constituent elements (shareholders, managers, etc.) is essential. Similarly, legal guarantees for foreign investment (particularly guarantees against appropriation or nationalization) are necessary before the process begins. Other legal reforms are important but can be put into place simultaneously with the implementation of the privatization process. A solid legal framework must precede privatization, but privatization should not wait until all relevant or helpful legislation is enacted.

7. Foreign aid, both in terms of money and in terms of expertise, can be extremely important in developing fledgling institutions that facilitate privatization. The SPA, the SSS, and the BSE all received such aid and were greatly benefited by it.

8. The ownership structures of state-owned businesses must be made clear before privatization begins. Lack of such clarity in Hungary cost that country millions of dollars in privatization revenues

Learning from the Hungarian Privatization Process

as a result of delays and disputes over ownership of state assets. Concomitantly, the question of how the shares in a newly transformed company should be distributed among various claimants would necessarily be decided. Strong and comprehensive national legislation, to the extent constitutionally permissible, would be best.

9. The government should be prepared for difficulties in valuation of state assets and should develop a strategy in advance for dealing with those difficulties to ensure that the valuation problems do not delay privatization efforts. It should consider not requiring a state-sponsored valuation against which to measure bids if it cannot be confident that such a valuation will provide information about true market value. Hungary wasted far too much time and effort on valuations that led to its insistence on prices having almost nothing to do with what a willing buyer would pay.

10. The government must be realistic about how low true market prices may be and be prepared to accept low prices in a weak market rather than insisting on supposed "fair" prices that may lead to the assets remaining in state hands and further deteriorating in value.

11. The government must also resign itself to substantial foreign participation (or even domination) in the privatization process unless it can count on strong domestic demand. If existing domestic demand is not strong enough, the government should begin taking steps early to boost that demand, and these steps should include an analysis of what is holding back domestic demand. In Hungary, lack of available credit, lack of investment funds in which to pool monies, and lack of experience as shareholders led to specific regulatory overtures directed at those problems. Each country's situation is unique and must be analyzed separately.

12. The benefits of selling assets rather than giving them away should not be overestimated. In particular, the government must be realistic about the possibility that privatization revenues will be low. Although ensuring that those to whom assets are distributed have both the expertise and incentive to maximize the use of those assets may militate in favor of sales rather than giveaways, certain long-term benefits accrue if assets are distributed domestically and quickly. These include the rapid creation of a private economy, the creation of a strong middle class with a stake in political and economic reform, and broad wealth dispersal through the country. In retrospect, insisting upon sales netted Hungary little revenue and probably

cost it substantially in terms of economic growth. A future comparative study of Hungary, Poland, and the Czech Republic may shed light on which of the two systems (giveaways or sales) is better.

13. If assets are to be given away, partial giveaways in the form of subsidized sales are probably more desirable than outright gifts. Subsidized sales ensure a partial economic stake in the enterprise for the owner and are self-selective in that only those who think themselves capable of managing assets will buy them. If shares are given away widely, an easily accessed secondary market for the shares should be facilitated to permit those shares to be transferred to those best able to manage the companies involved. Inflationary pressures resulting from the rapid creation of new wealth would have to be absorbed.

14. In the overall transformation of the economy from one that is primarily state controlled to one that is primarily privately owned, privatization is less important than other events such as the growth of the private economy (i.e., entrepreneurial activity) and nonprivatization direct investment and the decline of the state sector. Indeed, as the state sector falters, the transformation of the economy will take place without formal privatization so long as the state encourages the growth of the private economy. That is, as the state sector declines and the private sector grows, the percentage of the economy in private hands will grow even without privatization. This suggests that putting resources into encouraging entrepreneurial activity (e.g., by lowering tax rates) would be wiser than putting resources into privatization if the goal is to establish a private economy.

15. Rapid privatization is preferable to slow privatization. The most rapid pace sustainable by political realities should be pursued. The inevitable economic chaos and dislocations are better suffered quickly than drawn out over many years.

16. If privatizations occur through sales, competitive bidding should not be required. It appears attractive, and in some cases it leads to higher prices, but it should be used only in those cases in which the privatizing agency has identified a property that will be attractive to many bidders. For other companies, any reasonable bid should be accepted quickly. In particular, if a bidder (as opposed to the privatizing agency) identifies a target for privatization, that target should not be shopped to other bidders until bilateral negotiations have been exhausted. Otherwise potential bidders will be reluctant to sink costs into identifying possible targets.

17. Foreign ownership should not be a concern so long as it can be sold politically. The benefits of foreign business expertise are substantial, high levels of foreign ownership are common in many countries, and foreign purchases may be unavoidable for a country seeking to sell its assets rather than distribute them to the public.

18. The intertwined subjects of pace, foreign ownership, and sale vs. giveaways should be debated and policies on them formulated before privatization begins.

19. There should be a conscious mix of private placements and public offerings, whether assets are sold or given away. Private placements can be done more quickly, are attractive to investors that want controlling interests, and provide business expertise. Public offerings promote wide dispersal of ownership, ownership by small investors, and support for fledgling stock exchanges.

20. A minimum of assets should be retained in majority ownership by the state. Partial retention of former state firms for investment purposes may be wise if in fact those firms in private hands appear to be good investments. But except in vital strategic industries, government interests are adequately protected by regulation. The inefficiency of state ownership is the reason firms are being privatized in the first place. Thus state ownership should be minimized.

21. Strategies should be developed early for maximizing both domestic and foreign demand for state assets being sold. Domestically, that means solving whatever problems are inhibiting demand. Foreign demand can be enhanced by providing political stability, foreign investment guarantees and incentives, a convertible currency, and a legal and financial infrastructure to support the needs of foreign businesses.

22. Attempts should be made to ensure that legislation not directly related to privatization either promotes privatization or at least does not hinder it.

23. Legislation should be integrated to reduce policy conflicts and promote a single set of government policies, including privatization.

24. Regulatory regimes surrounding the privatization efforts (e.g., company law, securities law, tax law) should only be as complex as necessary and should avoid being overly bureaucratic. On the other hand, these laws should be enacted with sufficient care to avoid foreseeable problems.

Notes

1. Suzanne Loeffelholz, "Dealing on the Danube," *Financial World*, Mar. 6, 1990, 38.
2. "The Privatizing of Eastern Europe," *Institutional Investor* (April 1990), quotes German financier.
3. For my purposes, the unique characteristics of privatization in former Socialist countries suggested here are most important. For other aspects of Eastern European privatization that make those efforts different from other countries' experiences, see David Bartlett, "The Political Economy of Privatization: Property Reform and Democracy in Hungary," *East European Politics and Societies* 6 (Winter 1992): 73, 76–77.
4. Plantagenet Fry & Fiona Somerset Fry, *The History of Scotland* (Boston: Routledge & Kegan Paul, 1982), 73.
5. See E. I. DuPont de Nemours & Co. v. Byrns, 1 F.R.D. 34, 38 (S.D.N.Y. 1939); Richard A. Posner, *Cardozo: A Study in Reputation* (Chicago: University of Chicago Press, 1990), 127.
6. Yair Aharoni, "The United Kingdom: Transforming Attitudes," in *The Promise of Privatization: A Challenge for U.S. Policy*, ed. R. Vernon (New York: Council on Foreign Relations, 1988).
7. Michel Bauer, "The Politics of State-Directed Privatization: The Case of France," in *The Politics of Privatization in Western Europe*, ed. J. Vickers and V. Wright (London: F. Cass, 1989).
8. Wolfgang Muller, "Privatizing in a Corporatist Economy: The Politics of Privatization in Austria," in *The Politics of Privatization in Western Europe*, ed. J. Vickers and V. Wright (London: F. Cass, 1989).
9. For description and analysis of the Brazilian example, see, Ethan Kapstein, "Brazil: Continued State Dominance," in *The Promise of Privatization: A Challenge for U.S. Policy*, ed. R. Vernon (New York: Council on Foreign Relations, 1988). On Latin America generally, see Eliana Cordoso, "Privatisation Fever in Latin America," *Challenge* (Sept.-Oct. 1991): 35–41.
10. For a description of privatization attempts in Africa and other parts of the third world, see "Privatisation in Developing Countries," *Financial Times*, Aug. 2, 1985, I 110. See also "All Around the Globe, the Sale of the Century," *Washington Post*, Nov. 17, 1991, H1, which describes privatization efforts in several countries.
11. On the Philippine example, see Stephen Haggard, "The Philippines: Picking Up Pieces After Marcos," in *The Promise of Privatization: A Challenge for U.S. Policy*, ed. R. Vernon (1988), 91.

12. In the United States, which retained a relatively small state sector after World War II, the receding of statism took the form of deregulation rather than privatization.

13. See, for example, "Investors Cool to India Privatization," *International Herald Tribune*, Dec. 28, 1992, p. 8, col. 4, which describes disappointing investor response to the second phase of India's privatization program, begun in October 1992 and designed to narrow the Indian budget deficit.

14. See "Hungary: A Race Against Time," *Financial Times*, Oct. 30, 1991.

15. See "Proposal on Privatization Strategies in Hungary," *MTI Econews*, June 3, 1991.

16. See "In Eastern Europe, the Big Sell-Off Is Set to Begin," *Business Week*, Aug. 6, 1990, 42.

17. "Hungary's Big Sell-Off Hits Hard Times," *Financial Times*, May 21, 1992, I12. See also "Eastern Europe—Early Days," *Euromoney Supplement: The Global Sweep of Privatization* (July 1990): 2, which states "There is no precedent for the developments due to take place in eastern Europe"; "On the Back Burner," *Hungarian Observer* (Oct. 1990), states, "No other country in the world has set out to privatize state assets on this scale."

18. See, for example, Ziya Onis, "Privatization and the Logic of Coalition Building: A Comparative Analysis of State Divestiture in Turkey and the United Kingdom," *Comparative Political Studies* 24 (July 1991): 231.

19. David Bartlett, "The Political Economy of Privatization: Property Reform and Democracy in Hungary," *East European Politics and Societies* 6 (Winter 1992): 73, 74–75, states: "Whereas representative political institutions promoted the interests of private capital in the early developing countries, they create certain obstacles to privatization in late developing countries undergoing transitions from authoritarianism." One problem seen in developing countries that is being avoided in Hungary is the tendency for privatized companies to be turned over to private monopolists, who are no more likely to be efficient than the state monopolies that proceeded them. See "Privatisation in Developing Countries," *Financial Times*, Aug. 2, 1985, I10.

20. See infra Chapter Nine.

21. Emilia Sebok, "Speeding Up Privatization," *Hungarian Economy* 18, nos. 3-4 (1990): 10.

22. Yet other countries' experiences may prove more relevant in certain respects than Hungary's. For example, Albert Chernyshev, the Russian ambassador to Turkey, was quoted in late 1992 as suggesting Turkey's privatization program as an appealing model for Russia: "Turkey has undergone the first phases of establishing political democracy and a free market economy, and this has been done often in a crisis environment, which is also the case with the C.I.S. countries. This kind of model is intriguing, even for us. We are looking at the Turkish experience." "Across the Great Divide," *Time International*, Oct. 19, 1992, 31.

23. See infra Chapter Three.

24. "Eastern Europe—Early Days," *Euromoney Supplement: The Global Sweep of Privatization* (July, 1990): 2.

25. Suzanne Loeffelholz, "Dealing on the Danube," *Financial World*, Mar. 6, 1990, 38, 42.

26. "World Bank Loan to Hungary," *MTI Econews*, Feb. 15, 1992.

27. "Head of U.S. Investment Fund Criticizes Slow Pace of Hungarian Privatization," *BNA International Trade Reporter*, Feb. 19, 1992, 322.

28. Hungary is a parliamentary democracy. Parliamentary statutes are the most important sources of Hungarian law. Ministerial decrees generally provide regulatory detail. Court decisions are also important sources of law, but their relative importance is diminished somewhat by Hungary's history as a civil law country.
29. As we see below, problems occasionally arise when laws that seem not to deal with privatization nonetheless affect it. See infra pages 22–26 and accompanying text.
30. III HRLF 3-4 (1992). Hungarian statutes are published in session law form in Hungarian Rules of Law in Force, abbreviated HRLF. Each year begins a new volume, and the publications come out in numbered form approximately biweekly. Each edition is published in Hungarian with translations in German and English.
31. Act VI of 1959 on the Civil Code of the Peoples' Republic of Hungary, chap. 46, I HRLF 21-24 (1990).
32. Company Law of 1875.
33. Interview with Dr. Andras Kisfaludi, 1991.
34. Act VI of 1988 on Business Organizations, preamble. "The aim of this law is to enhance, by establishing a modern legal framework, the improvement of the income-generating capability of the economy, the development of market-related production and marketing cooperation, the flow of capital, and the direct presence of foreign capital in our economy."
35. Ibid., chap. 1.
36. Ibid., art. 9.
37. Ibid., arts. 19, 23, 24.
38. Ibid., chap. 2.
39. Ibid., art. 20.
40. Act LXV of 1991, art. 5.
41. Act VI of 1988 on Business Organizations, part II.
42. Ibid., art. 55(1).
43. Ibid., art. 63.
44. Ibid., art. 64.
45. Ibid., art. 66.
46. Ibid.
47. Ibid., art. 75.
48. Ibid.
49. Ibid., art. 83.
50. Ibid., art. 94.
51. Ibid., art. 97.
52. Ibid., art. 98.
53. Ibid., arts. 97(4), 98.
54. Ibid., art. 103.
55. Ibid.
56. Ibid., arts. 127, 132.
57. Ibid., art. 127.
58. Ibid., art. 133(1).
59. Ibid., art. 134(1).
60. Ibid., art. 142.
61. Ibid., art. 140.
62. Ibid., art. 142.
63. Arts. 155, 156, 158, 167. Throughout this work, Hungarian forints are converted to U.S. dollars at the rate of 75 forints per dollar. This is an approximation.

During the period under study, the Hungarian forint has decreased in value relative to the dollar from approximately 68 forints per dollar in 1990 to approximately 82 forints per dollar in 1992. The consistent use of the figure 75 forints per dollar is meant to approximate the conversion to provide a perspective to those readers not familiar with forint values. The approximate nature of the conversions is not significant to the points raised.

64. Ibid., art. 156.
65. Ibid., art. 171.
66. Ibid., arts. 183-96.
67. Ibid., arts. 197-207.
68. Ibid., arts. 209-14.
69. Ibid., art. 232(1).
70. Ibid., art. 233.
71. Ibid., art. 251(1).
72. Ibid., arts. 277-84.
73. Ibid., arts. 285-90.
74. Ibid., arts. 291-96.
75. I HRLF 1 (1990), as amended II HRLF 5 (1991).
76. Act VI of 1988 on Business Organizations, art. 14.
77. Act XXIV of 1988 on Investments of Foreigners in Hungary, art. 1.
78. Ibid., art. 4(2), I HRLF 1 (1990).
79. Ibid., art. 9(2).
80. Act XCVIII of 1990. See Act XXIV of 1990 on Foreign Investments, codified as amended II HRLF 5 (1991).
81. Ibid., art. 16.
82. Ibid.
83. Ibid., art. 16.
84. Ibid., art. 18.
85. See infra Chapter Seven.
86. I HRLF 25 (1990).
87. Ibid., art. 2(1)(c).
88. Ibid., arts. 5, 6.
89. Ibid., art. 7.
90. Ibid., arts. 7, 16, 17.
91. Ibid., art. 17/A.
92. Ibid., art. 14.
93. Ibid., art. 17(3).
94. Ibid., arts. 8, 9.
95. See infra Chapter Four.
96. I HRLF 8 (1990).
97. Interview with Dr. Tamas Rusznak, 1981.
98. Ministry of Finance, "Preamble to the Act on Securities and the Stock Exchange," *Public Finance in Hungary* 64 (990): 3, 6-7.
99. Interview with Kalman Debreczeni, January 1991. Debreczeni, chairman of an investment banking and brokerage firm, describes the act as "awfully bureaucratic."
100. Act VI of 1990 on Securities and the Stock Exchange, secs. 1-3.
101. Interview with Dr. Tamas Rusznak, Deputy Head of State Banking Supervision, January 1991.
102. Interview with Zsuzsanna Zatik, Girozentrale Investment Ltd., January 1991.

Notes

103. Act VI of 1990 on Securities and the Stock Exchange, secs. 4-7.
104. Ibid., secs. 8-9.
105. Interview with Dr. Zoltan Pacsi, January 1991. One of the principal drafters of the act agreed with this interpretation of the statute. Interview with Dr. Tamas Rusznak, January 1991.
106. Act VI of 1990 on Securities and the Stock Exchange, secs. 10-16.
107. Ibid., secs. 23-32.
108. Ibid., sec. 24.
109. Ibid., secs. 33-34.
110. Ibid., secs. 35-41.
111. Act LXXXVI of 1990 on the Prohibition of Unfair Market Practices, II HRLF 5 (1991).
112. Act VI of 1990 on Securities and the Stock Exchange, secs. 42-74.
113. Ibid., secs. 45-49, 65-67.
114. Ibid., sec. 68.
115. Ibid., secs. 75-77.
116. Ibid., secs. 78-79.
117. Interview with Mr. Theodore Boone, January 1991.
118. Ibid.
119. Act VI of 1990 on Securities and the Stock Exchange, sec. 81.
120. Ibid.
121. Ibid., sec. 84.
122. Interview with Dr. Zsuzsanna Zatik, January 1991; Interview with Mr. Theodore Boone, January 1991.
123. Interview with Dr. Tamas Rusznak, January 1991.
124. Some assistance was provided by the Ministry of Finance, which published a pamphlet describing and interpreting the various sections of the act one by one. Hungarian Ministry of Finance, "Preamble to the Act on Securities and the Stock Exchange," *Public Finance in Hungary* 64 (1990):3. This pamphlet, however, was not an official interpretation of the act.
125. I HRLF 9 (1990).
126. Ibid., sec. 1.
127. Ibid., sec. 2.
128. Ibid., sec. 3.
129. Ibid., secs. 4-5.
130. Ibid., secs. 7-9.
131. Ibid., secs. 10-19.
132. Ibid., secs. 20-22.
133. Ibid., sec. 23.
134. Ibid., secs. 28-33.
135. See infra Chapter Four.
136. See infra Chapter Nine.
137. See infra Chapter Five.
138. I HRLF 25 (1990).
139. See infra Chapter Three.
140. Act VIII of 1990 on the Protection of Property Entrusted to Enterprises of the State, secs. 1-3.
141. Ibid., sec. 1.
142. I HRLF 25 (1990).
143. See infra Chapter Four.

144. Act LXXIV of 1990 on the Privatization (Alienation, Utilization) of State-Owned Companies Engaged in Retail Trade, Catering, and Consumer Services, secs. 2, 16.
145. Ibid., sec. 8; Act V of 1990 on Private Enterprise, sec. 3, 1 HRLF 25 (1990).
146. Act LXXIV of 1990 on the Privatization (Alienation, Utilization) of State-Owned Companies Engaged in Retail Trade, Catering, and Consumer Services, sec. 9.
147. Ibid., sec. 10.
148. II HRLF 15 (1991).
149. "The Privatisation Dilemma," *Euromoney* (Sept. 1990): 145-46.
150. Act XVIII of 1991 on Accounting, preamble.
151. See infra Chapter Ten.
152. III HRLF 1 (1991).
153. II HRLF 16 (1991).
154. "Hungary's Industry Up for Sale," *Business Law Brief, Financial Times*, (Oct. 1990).
155. "Hungary's Industry Up for Sale," *Business Law Brief, Financial Times*, (Oct. 1990), secs. 2-12. The 1949 date chosen was the date of the Communist consolidation of power. Importantly, the fascist confiscation of Jewish property before and during the Second World War was not covered by the Compensation Act, although the government promised a second statute to cover the period from 1939 to 1949. See sec. 1(3) also.
156. The Danubius and Pick Salami offerings of 1992 were typical in this regard. See infra Chapter Seven.
157. Act XXV of 1991 on Partial Compensation for Damages Unlawfully Caused by the State to Properties Owned by Citizens in the Interest of Settling Ownership Relations, secs. 13-28.
158. The total compensation permitted for any one property and for any one former owner was set at 5 million HuF, or approximately $67,000, but a sliding scale ensured that any claimed amount in excess of 500,000 HuF would be compensated at a 10 percent level. Act XXV of 1991 on Partial Compensation for Damages Unlawfully Caused by the State to Properties Owned by Citizens in the Interest of Settling Ownership Relations, sec. 4.
159. II HRLF 23 (1991).
160. See, for example, Law Decree 11 of 1986 on Liquidation Procedure; Decree No. 79/1988 on State Liquidation.
161. See infra Chapter Five.
162. III HRLF 18 (1992).
163. Ibid., secs. 5-8, 14-15.
164. See infra Chapter Four.
165. See infra Chapter Six.
166. Act LX of 1991 on the National Bank of Hungary, Art. 84(1), discussed in "Prodded by New Law, Hungarian Banks Edge Away From State Control Toward Privatization," *EBRD Watch*, 2, no. 32 (Aug. 31, 1992):3.
167. III HRLF 2 (1992).
168. Ibid., preamble.
169. Ibid., secs. 5, 7.
170. Ibid., secs. 29-38.
171. Ibid., sec. 9.
172. Ibid., sec. 12.

173. Ibid., sec. 40.
174. Ibid., secs. 14-19.
175. Ibid., secs. 23-28.
176. Ibid., sec. 52.
177. II HRLF 24 (1991).
178. Ibid., arts. 18(1), 97(3), discussed in "Privatization of Hungarian Banks: State Must Reduce Ownership Share by Law," *MTI Econews*, Nov. 26, 1991; "Privatization of Large Banks," *Hungarian Observer* (June 1992).
179. Act LXIX of 1991 on Banks and Banking Activities, art. 97(4), discussed in "Permanent State Ownership in Commercial Banks," *MTI Econews*, Aug. 31, 1992.
180. Law LXIX of 1991 on Banks and Banking Activities, Arts. 21-23, 98-101, discussed in "Privatization of Hungarian Commercial Banks," *MTI Econews*, Jan. 14, 1992.
181. The Hungarian term for this organ: *Allami Ertekpapir Felugyelet* is variously translated as "State Securities Supervisory Board," "State Securities Supervision," or "State Securities Directorate." The translation used here is one of several that may be encountered by the reader.
182. Act VI of 1990 on Securities and the Stock Exchange, secs. 4-22.
183. Interview with Dr. Zoltan Pacsi, January 1991; interview with Antonia Szabo, January 1991; interview with Antonia Szabo, March 1993.
184. Ibid.
185. Act LX of 1991 on the National Bank of Hungary, III HRLF 18 (1992); Act LXIX of 1991 on Banks and Banking Activities, II HRLF 24 (1991).
186. Act VII of 1990 on Foundation of State Property Agency with the Purpose of the Management and Utilization of Property Pertaining to This, I HRLF 9 (1990); Act VIII of 1990 on the Protection of Property Entrusted to Enterprises of the State, I HRLF 25 (1990); Act LXXIV of 1990 on the Privatization (Alienation, Utilization) of State-Owned Companies Engaged in Retail Trade, Catering, and Consumer Services, I HRLF 25 (1990).
187. David Bartlett, "The Political Economy of Privatization: Property Reform and Democracy in Hungary," *East European Politics and Societies* 6 (Winter 1992):73, 94-97.
188. "Hungary Hurries to Privatisation," *Times* (London) (June 5, 1990), Business Section, quotes Istvan Tompe, managing director of SPA.
189. See Stark, "Privatization in Hungary: From Plan to Market or From Plan to Clan?" *East European Politics and Societies* 4 (Fall 1990):351, 356-57.
190. See infra pages 47-52.
191. See, for example, Gyorgy Matolcsy, "Privatization: Hungary," *East European Economics* 30 (Fall 1991): 49, 52. Matolcsy argues that economic performance data from 1990 indicate that, notwithstanding the problems surrounding spontaneous privatization, the process was working to boost the economy. Matolcsy points, however, to macroeconomic data such as the aggregate balance of trade and current accounts for Hungary. Given the relatively small impact of spontaneous privatization on the Hungarian economy in 1990, however, he clearly overstates the case.
192. "In Eastern Europe, The Big Sell-Off Is Set to Begin," *Business Week*, Aug. 6, 1990, 42.
193. Ibid.
194. See "Legislation on Privatization and Acquisitions," *Doing Business with Eastern Europe*, Aug. 1, 1991.

195. See "Legislation on Privatization and Acquisitions," *Doing Business With Eastern Europe*, Aug. 1, 1991.
196. "Acquisitions in the East: Europe's New M & A Frontier," *M & A Europe* (Nov.-Dec. 1990).
197. See "Privatisation in Eastern Europe," *Economist*, April 14, 1990, 20.
198. Ibid.
199. "Acquisitions in the East: Europe's New M & A Frontier," *M & A Europe* (Nov.-Dec. 1990).
200. Ibid.
201. See "The Privatisation Dilemma," *Euromoney* (Sept. 1990): 145–46.
202. "East Europe's Sale of the Century," *New York Times*, May 22, 1990, D1.
203. Suzanne Loeffelholz, "Dealing on the Danube," *Financial World*, Mar. 6, 1990, 39.
204. "Unhappy Hunting Ground," *International Management* (Sept. 1990): 61.
205. "The Privatizing of Eastern Europe," *Institutional Investor* (April 1990):171, 174.
206. Stark, "Privatization in Hungary: From Plan to Market or From Plan to Clan?" *East European Politics and Societies* 4 (Fall 1990): 351, 357.
207. Ibid., 358.
208. See Stark, "Privatization in Hungary: From Plan to Market or from Plan to Clan?" *East European Politics and Societies* 4 (Fall 1990): 351, 365–66.
209. See "The Privatisation Dilemma," *Euromoney* (Sept. 1990): 145–46.
210. The privatization debate preceding the 1990 election is chronicled in David Bartlett, "The Political Economy of Privatization: Property Reform and Democracy in Hungary," *East European Politics and Societies* 6 (Winter 1992): 73.
211. See Stark, "Privatization in Hungary: From Plan to Market or From Plan to Clan?" *East European Politics and Societies* 4 (Fall 1990): 351, 366.
212. "Hungary Spurs Privatisations as Economic Crisis Looms," *Reuters Library Report*, Nov. 4, 1990.
213. Along with those mentioned here, an important element of the list was Act VI of 1990 on Securities and the Stock Exchange, which, *inter alia*, made possible the reopening of the Budapest Stock Exchange and created the State Securities Supervision.
214. State Property Agency, "Information Booklet (1991).
215. See "Acquisitions in the East: Europe's New M & A Frontier," *M & A Europe* (Nov.-Dec. 1990).
216. "Privatization in Hungary: Official Lists Not Needed," *Business Eastern Europe*, Aug. 13, 1990, 267.
217. "SPA Report on 1991," *MTI Econews,* July 9, 1992.
218. As discussed later in this chapter, the task of managing state assets that would not be privatized immediately, including retained minority stakes in privatized firms, devolved to a separate organization in late 1992.
219. "Hungary Seeks UK Help on Privatisation," *Financial Times*, Mar. 8, 1990, sec. 1, p. 3.
220. See David Bartlett, "The Political Economy of Privatization: Property Reform and Democracy in Hungary," *East European Politics and Societies* 6 (Winter 1992): 73, 106.
221. State Property Agency, "Information Booklet (1991).
222. See "Privatization in Hungary: The Art of the Possible," *International Financial Law Review*, 10 (April 1991): 32.

223. Interviews at State Property Agency, November 1, 1992.
224. See infra Chapter Nine.
225. "Hungary's Industry Up for Sale," *Business Law Brief, Financial Times* (Oct. 1990).
226. Ibid.
227. See "Acquisitions in the East: Europe's New M & A Frontier," *M & A Europe* (Nov.-Dec. 1990).
228. "Hungary Spurs Privatisations As Economic Crisis Looms," *Reuters Library Report*, Nov. 4, 1990, quotes Les Bonnay of Price Waterhouse.
229. "Hungary Puts Speed Above Control in Privatization," *Reuters Money Report*, Jan. 2, 1992.
230. With only approximately 10 million people and a language that is not even in the Indo-European language family (Sanskrit is closer linguistically to German, English, Russian, French, and Italian than Hungarian is), Hungarians could not reasonably expect all foreigners to deal with them in their native tongue.
231. Interviews at the State Property Agency, Nov. 1, 1992.
232. "Hungary's Dance with Capitalism," *Los Angeles Times*, Aug. 27, 1989, Business section, p. 1, col. 5, quotes Zsigmond Jarai, deputy finance minister.
233. Nigel Ash, "The Gearbox Needs an Overhaul," *Euromoney Supplement* (Apr. 1992): 41, 47.
234. "The Privatizing of Eastern Europe," *Institutional Investor* (April 1990): 171, 174.
235. See Stark, "Privatization in Hungary: From Plan to Market or from Plan to Clan?" *East European Politics and Societies* 4 (Fall 1990): 351, 359.
236. Gerd Schwartz, "Privatization: Possible Lessons from the Hungarian Case," *World Development* 19 (1991): 1731-1732, puts the number of SOEs at more than 2000; "Hungary's Big Sell-Off Hits Hard Times," *Financial Times*, May 21, 1992, I12, puts number at 2000; "SPA Report on 1991, *MTI Econews*, July 9, 1992, states that the SPA reported 2244 SOEs as of Jan. 1, 1991; Sulkowski, Glick, and Richter, "Privatization in Hungary: The Art of the Possible," *International Financial Law Review* 10 (April 1991): 32, puts the number of SOEs at 2400.
237. Russell Johnson, "Privatization in Hungary Is Creating Investment Opportunities for U.S. Firms," *Business America*, Jan. 14, 1991, 112, estimates that 90 percent of the Hungarian economy was in state hands in early 1990s; Sulkowski, Glick, and Richter, "Privatization in Hungary: The Art of the Possible," *International Financial Law Review* 10 (April 1991): 32-33, reports that state-owned enterprises account for 85-90 percent of industrial output; Suzanne Loeffelholz, "Dealing on the Danube," *Financial World*, Mar. 6, 1990, reports that state-owned enterprises account for 65 percent of production but 95 percent of economically active assets; "Hungary Hurries to Privatisation," *Times* (London), June 5, 1990, Business Section, quotes Istvan Trompe, managing director of SPA, as estimating that 85 percent of Hungarian economy is in state hands; Emilia Sebok, "Speeding Up Privatization," *Hungarian Economy* 18, nos. 3-4 (1990): 10, estimates that 90 percent of enterprises in Hungary are state owned.
238. "Hungary Privatisation Programme Hits Crucial Phase," *Reuters Money Report,* Dec. 20, 1991, reports estimate of Peter Rajcsanyi, former SPA head of state-directed privatization.
239. Bela Csikos Nagy, "Privatization in a Post-Communist Society—The Case of Hungary," *Hungarian Business Herald* 2 (1991): 36-37.
240. The Soviet Union represented approximately 30 percent of Hungary's

foreign trade in 1989 and 14 percent in 1990. See "Hungary: A Giant Step Ahead," *Business Week,* Apr. 15, 1991, 58. COMECON was formally disbanded in 1991.

241. David M. Newberry, "Reform in Hungary—Sequencing and Privatisation," *European Economic Review* 35 (1991): 571, 578 n.2.

242. Bela Csikos Nagy, "Privatization in a Post-Communist Society—The Case of Hungary," *Hungarian Bussiness Herald* 2 (1991): 36–37.

243. See infra Chapter Ten.

244. "The Privatizing of Eastern Europe," *Institutional Investor* (April 1990); "East Europe's Sale of the Century," *New York Times,* May 22, 1990, D1, col. 3.

245. "Quoted in "The Privatizing of Eastern Europe," *Institutional Investor* (April 1990): 173.

246. "Privatisation in Eastern Europe," *Economist,* April 14, 1990. See also "East Europe's Sale of the Century," *New York Times,* May 22, 1990, D1, col. 3, quotes former SPA managing director Istvan Tompe as saying that economic revival is one goal of privatization; "Hungary Pressed to Privatise," *East European Markets,* April 17, 1992, suggests that Hungary's ability to meet its energy needs depends on its privatization of its moribund energy assets.

247. "Hungary," *Internatinal Reports,* Sept. 7, 1990, recounts statement of head of Hungarian National Bank.

248. "The Privatizing of Eastern Europe," *Institutional Investor* (April 1990); Suzanne Loeffelholz, "Dealing on the Danube," *Financial World,* Mar. 6, 1990, 42.

249. Gerd Schwartz, "Privatization: Possible Lessons from the Hungarian Case," *World Development* 19 (1991): 1731. In October 1992, the 1993 deficit was projected at $2.8 billion. For a country of only 10 million inhabitants with a gross national product nearly at Third World levels, the deficit was alarming.

250. Professor Sachs' ideas set forth in "Sachs, the Property Distributor," *Privat Profit* (Aug. 1991) produced a sharp rebuke from an Hungarian author in "An Open Letter About Property," *Privat Profit* (Aug. 1991).

251. David M. Newberry, "Reform in Hungary—Sequencing and Privatisation," *European Economic Review* 35 (1991): 571, 575, 578–79.

252. O. Blanchard and R. Layard, "Economic Change in Poland, Discussion Paper No. 424, Center for Economic Policy Research, London, 1990.

253. See, for example, Stark, "Privatization in Hungary: From Plan to Market or from Plan to Clan?" *East European Politics and Societies* 4 (Fall 1990) 351, 386–88; Gerd Schwartz, "Privatization: Possible Lessons from the Hungarian Case," *World Development* 19 (1991): 1731, 1734.

254. Suzanne Loeffelholz, "Dealing on the Danube," *Financial World,* Mar. 6, 1990, 42.

255. Bela Csikos Nagy, "Privatization in a Post-Communist Society—The Case of Hungary," *Hungarian Business Herald* (1991): 36, 39.

256. "Hungary's Industry Up for Sale," *Business Law Brief, Financial Times* (Oct. 1990).

257. See, for example, Stark, "Privatization in Hungary: From Plan to Market or from Plan to Clan?" *East European Politics and Societies* (Fall 1990): 351, 382–85.

258. David Bartlett, "The Political Economy of Privatization: Property Reform and Democracy in Hungary," *East European Politics and Societies* 6 (Winter 1992): 73, 105.

259. Ibid.

260. Bela Csikos Nagy, "Privatization in a Post-Communist Society — The Case of Hungary," *Hungarian Business Herald* 2 (1991): 36, 39.
261. "The Privatizing of Eastern Europe," *Institutional Investor* (April 1990): 171, 178.
262. Suzanne Loeffelholz, "Dealing on the Danube," *Financial World*, Mar. 6, 1990, 42.
263. Stark, "Privatization in Hungary: From Plan to Market or from Plan to Clan?" *East European Politics and Societies* 4 (Fall 1990): 351, 373.
264. Ibid.
265. See infra chapters Five and Seven.
266. See supra pages 38–39.
267. For a description of the Czech and Polish programs, see David Bartlett, "The Political Economy of Privatization: Property Reform and Democracy in Hungary," *East European Politics and Societies* 6 (Winter 1992): 73, 82-90; Vratislav Pechota, "Privatization and Foreign Investment in Czechoslovakia: The Legal Dimension," *Vand. J. Transnational L.* 24 (1991): 308; Zbigniew Fallenbuchl, "Polish Privatization Policy," *Comparative Economic Studies* 33 (1991): 65. For a comparison of the Hungarian and Czech programs, see "Czechoslovakia, Hungary Choose Differing Routes to Reach Same Goal — Privatization," *EBRD Watch* 2, no. 19 (May 18, 1992): 4.
268. One might argue, however, that giving state assets away, as opposed to selling them, would not revitalize the economy as quickly. A giveaway might have serious adverse effects on inflation, and the recipients might not be as efficient users as those confident enough in their own abilities to be enticed to part with hard-earned money.
269. See supra Chapter Two.
270. Suzanne Loeffelholz, "Dealing on the Danube," *Financial World*, Mar. 6, 1990, 39; "The Privatizing of Eastern Europe," *Institutional Investor* (April 1990): 171, 173; "Survey of Privatisation in Eastern Europe," *Financial Times*, July 3, 1992; "Off on the Wrong Foot," *Hungarian Observer* (Jan. 1991).
271. "Off on the Wrong Foot," *Hungarian Observer* (Jan. 1991).
272. General Electric made its ownership of Tungsram the focus of a massive advertising campaign, lauding its international cooperation in aid of the former subjects of Communist rule and its breaking of a significant business barrier.
273. Interviews at CA-BB, January 1991; "The Privatisation Dilemma," *Euromoney* (Sept. 1990): 145–46.
274. "The Privatisation Dilemma," *Euromoney* (Sept. 1990): 145–46; interviews at CA-BB, January 1991.
275. "Hungary Hurries to Privatisation," *Times* (London), June 5, 1990, Business Section.
276. Ibid.
277. "IBUSZ Papers for Compensation Coupons," *MTI Econews*, June 22, 1992.
278. "East Europe's Sale of the Century," *New York Times*, May 22, 1990, D1, col. 3.
279. Gerd Schwartz, "Privatization: Possible Lessons from the Hungarian Case," *World Development* 19 (1991): 1731, 1733.
280. "Privatisation Row Erupts in Hungary," *Independent*, July 6, 1990, 21.
281. "Hungarian State Body Attacked on Privatisation," *Financial Times*, July 6, 1990, 12.

282. "Hungarian State Body Attacked on Privatisation," *Financial Time*, July 6, 1990, I2.
"Privatisation Row Erupts in Hungary," *Independent*, July 6, 1990, 21.
283. "Hungarian State Body Attacked on Privatisation," *Financial Times*, July 6, 1990, I2.
284. Ibid.
285. "Dispute in Hungary Privatization," *Business Eastern Europe* July 16, 1990, 235.
286. "Hungary Hurries to Privatisation," *Times* (London), June 5, 1990, Business Section.
287. "The Painful Road to Privatisation," *Financial Times*, Nov. 5, 1991, 129.
288. "Hungary Replaces Criticized Privatization Chief," *Reuters Library Report*, July 26, 1990.
289. "Brake on Hungary's Privatisation Drive," *Times* (London), July 30, 1990, Business Section.
290. "State Property Agency to Sell Its IBUSZ Shares," *MTI Econews*, Nov. 21, 1991.
291. Ibid.
292. Ibid.
293. Indeed, IBUSZ's Articles of Association provided that the state would retain 51 percent of voting rights of shareholders so long as it held at least 33 percent of the issued shares. See State Property Agency, *First Privatization Program* (1990). The state was clearly ambivalent about the merits of private ownership.
294. "Way Cleared for IBUSZ Privatization," *BBC Summary of World Broadcasts*, Feb. 29, 1992.
295. "Permanent State Ownership in Commercial Banks," *MTI Econews*, Aug. 31, 1992.
296. "IBUSZ Papers for Compensation Coupons," *MTI Econews*, June 22, 1992.
297. Ibid., The subject of compensation coupons is taken up at length in a subsequent chapter.
298. See supra pages 42–43.
299. Gyorgy Matolcsy, "Privatization: Hungary," *East European Economics* 30 (Fall 1991): 49, 51.
300. Ibid.
301. "Legislation on Privatization and Acquisitions," *Doing Business With Eastern Europe*, Aug. 1, 1991.
302. "Privatization in Eastern Europe," *Economist*, Apr. 14, 1990, 20.
303. Gyorgy Matolcsy, "Privatization: Hungary," *East European Economics* 30 (Fall 1991): 49, 51.
304. "Legislation on Privatization and Acquisitions," *Doing Business with Eastern Europe*, Aug. 1, 1991.
305. "Privatization in Eastern Europe," *Economist*, April 14, 1990.
306. The problem of foreign dominance of the privatization process is taken up later in this chapter.
307. "Privatisation in Eastern Europe," *Economist*, Apr. 14, 1990, 20; "The Privatization of Eastern Europe," *Institutional Investor* (Apr. 1990): 171, 174, quotes Zsigmond Jarai.
308. "Eastern Europe—Early Days," *Euromoney Supplement: The Global Sweep of Privatization* (July 1990): 2; "Privatisation in Eastern Europe," *Economist*, Apr. 14, 1990, 20; "Acquisitions in Eastern Europe: Europe's New M & A Frontier," *M & A Europe* (Nov.-Dec. 1990).

309. "Acquisitions in Eastern Europe: Europe's New M & A Frontier," *M & A Europe* (Nov.-Dec. 1990); "In Eastern Europe, the Big Sell-Off Is Set to Begin," *Business Week*, Aug. 6, 1990, 42.
310. See supra pages 31-32, 42-43.
311. "The Privatization of Eastern Europe," *Institutional Investor* (Apr. 1990): 171, 173; "Eastern Europe—Early Days," *Euromoney Supplement: The Global Sweep of Privatization* (July 1990): 2.
312. "Calling for the Next Stage in Property Rights; An Interview with Dr. Lajos Csepi," *Privat Profit* (Sept. 1991).
313. Gyrogy Matolcsy, "Privatization: Hungary," *Eastern European Economics* 30 (Fall 1991): 49, 54.
314. Ibid.
315. "In Eastern Europe, the Big Sell-Off Is Set to Begin," *Business Week*, Aug. 6, 1990, 42.
316. David Bartlett, "The Political Economy of Privatization: Property Reform and Democracy in Hungary," *East European Politics and Societies* 6 (Winter 1992): 73, 104.
317. Emilia Sebok, "Speeding Up Privatization," *Hungarian Economy* 18, nos. 3-4 (1990): 10.
318. Russell Johnson, "Privatization in Hungary Is Creating Investment Opportunities for U.S. Firms," *Business America*, Jan. 14, 1991, 112.
319. Suzanne Loeffelholz, "Dealing on the Danube," *Financial World*, Mar. 6, 1990, 38.
320. "The Privatizing of Eastern Europe," *Institutional Investor* (Apr. 1990): 171, 174.
321. "Hungary: A Giant Step Ahead," *Business Week*, Apr. 15, 1991, 58.
322. Ibid.
323. "Head of U.S. Investment Fund Criticises Slow Pace of Hungarian Privatization," *(BNA International Trade Reporter,* Feb. 19, 1992, 322.
324. See supra pages 31-32.
325. "Eastern Europe—Early Days," *Euromoney Supplement: The Global Sweep of Privatization* (July 1990): 2.
326. "Managing the Managers, *Reason* (Mar. 1991): 48.
327. "In Eastern Europe, the Big Sell-Off Is Set to Begin," *Business Week*, Aug. 6, 1990, 42.
328. David Bartlett, "The Political Economy of Privatization: Property Reform and Democracy in Hungary," *East European Politics and Societies* 6 (Winter 1992): 73, 105.
329. Suzanne Loeffelholz, "Dealing on the Danube," *Financial World*, Mar. 6, 1990, 3p.
330. "Eastern Europe—Early Days," *Euromoney Supplement: The Global Sweep of Privatization* (July 1990): 20.
331. Gyorgy Matolcsy, "Privatization: Hungary," *Eastern European Economics* 30 (Fall 1991): 49, 51.
332. Sulkowski, Glick, and Richter, "Privatization in Hungary: The Art of the Possible," *International Financial Law Review* 10 (Apr. 1991):32; Gyorg Matolcsy, "Privatization: Hungary," *Eastern European Economics* 30 (Fall 1991) L:49, 52.
333. Stark, "Privatization in Hungary: From Plan to Market or from Plan to Clan?" *East European Politics and Societies* 4 (Fall 1990): 351, 367-69.
334. "Calling for the Next Stage in Property Rights; An Interview with Dr.

Lajos Csepi," *Privat Profit* (Sept. 1991). The fact that the spontaneous privatization route was the most popular with investors led the SPA to promise in 1991 it would try to simplify its other programs. See "Legislation on Privatization and Acquisitions," *Doing Business with Eastern Europe*, Aug. 1, 1991.

335. "Legislation on Privatization and Acquisitions," *Doing Business with Eastern Europe*, Aug. 1, 1991.

336. Ibid.

337. Gerd Schwartz, "Privatization: Possible Lessons from the Hungarian Case," *World Development* 19 (1991): 1731, 1733.

338. Gyorgy Matolcsy, "Privatization: Hungary," *Eastern European Economics* 30 (Fall 1991): 49, 52.

339. "Calling for the Next Stage in Property Rights; An Interview with Dr. Lajos Csepi," *Privat Profit* (Sept. 1991); "Too Many Firms, Too Few Buyers," *Economist*, Sept. 21, 1991, 14.

340. "Calling for the Next Stage in Property Rights; An Interview with Dr. Lajos Csepi," *Privat Profit* (Sept. 1991).

341. Gyorgy Matolcsy, "Privatization: Hungary," *Eastern European Economics* 30 (Fall 1991): 49, 53. This was to distinguish the program from the other programs for privatization that put the SPA in a passive position of passing on proposals initiated by either the companies themselves (spontaneous privatization) or investors (investor-initiated privatization).

342. "Hungarian Companies on the Block Are Named," *Financial Times*, July 13, 1990, I25; "Privatization in Hungary: Official Lists Not Needed," *Business Eastern Europe*, Aug. 13, 1990, 267.

343. "Hungary's Industry Up for Sale," *Business Law Brief, Financial Times* (Oct. 1990).

344. State Property Agency, "The State Property Agency (SPA) Announces the First Privatization Program (FPP) (Sept. 1990).

345. Gerd Schwartz, "Privatization: Possible Lessons From the Hungarian Case," *World Development* 19 (1991): 1731.

346. "Hungary Releases Its Privatisation Short-List," *Financial Times*, Sept. 15, 1990, I3.

347. Source: State Property Agency, "First Privatization Program (1990). Financial data are as of December 31, 1989, and are calculated under the old Hungarian accounting system, which in many respects is not consistent with that used in Western countries. In addition, performance is distorted in command economies.

348. David Fairlamb, "The Privatizing of Eastern Europe," *Institutional Investor* (Apr. 1990): 171, 178, reports criticism that second-rate companies needed the help more and that the good ones were sound investments for the state.

349. See "East Europe's Sale of the Century," *New York Times*, May 22, 1990, D1, col. 3; David Fairlamb, "The Privatizing of Eastern Europe," *Institutional Investor* (Apr. 1990): 171.

350. State Property Agency, "The State Property Agency (SPA) Announces the First Privatization Program (FPP) (Sept. 1990).

351. Ibid.

352. Widespread use of the BSE was unlikely; it was simply too small in its infancy to handle the larger flotations and was not interested in having poor equities dumped on it. See "Industry Minister Interviewed on Privatisation," BBC, June 23, 1990.

353. State Property Agency, "The State Property Agency (SPA) Announces the First Privatization Program (FPP) (Sept. 1990).
354. Ibid.
355. See Emilia Sebok, "Speeding Up Privatization," *Hungarian Economy* 18, nos. 3–4 (1990): 10.
356. State Property Agency, "The State Property Agency (SPA) Announces the First Privatization Program (FPP)" (Sept. 1990).
357. "Hungary Unveils First Phase of Major Privatization Program," *BNA Daily Report for Executives*, Sept. 18, 1990.
358. State Property Agency, "The State Property Agency (SPA) Announces the First Privatization Program (FPP)" (Sept. 1990).
359. Detail, quality, and depth of analysis of the issues facing the company and ideas on privatization options: 30 points
  Methodology, approach, and timetable for privatization proposed: 20 points
  Qualifications of bidder(s): 25 points
  Proposed fees: 25 points
360. State Property Agency, "The First Privatization Program (1990), attachment 3.
361. David Fairlamb, "The Privatizing of Eastern Europe," *Institutional Investor* (April 1990): 171.
362. Ibid.
363. "Too Many Firms, Too Few Buyers," *Economist*, Sept. 21, 1991, 14.
364. "First Privatization Programme Dragging On," *MTI Econews*, Aug. 27, 1991.
365. "Too Many Firms, Too Few Buyers," *Economist*, Sept. 21, 1991, 14.
366. "First Privatization Programme Dragging On," *MTI Econews*, Aug. 27, 1991.
367. Gyorgy Matolcsy, "Privatization: Hungary," *East European Economics* 30 (Fall 1991): 49, 53.
368. "Too Many Firms, Too Few Buyers," *Economist*, Sept. 21, 1991, 14.
369. "Slow Progress in Hungarian Privatisation," Agence France Presse, Oct. 15, 1991.
370. "Hungarians Need Bigger Role in Privatisation—SPA," *Reuters Money Report*, Jan. 13, 1992.
371. Ibid.
372. "Hungary's Big Sell-Off Hits Hard Times," *Financial Times*, May 21, 1992, 12.
373. Ibid.
374. Ibid.
375. "The Privatisation Dilemma," *Euromoney* (Sept. 1990): 145–46
376. Ibid.
377. Managing Director Lajos Csepi conceded, "It's obvious the bureaucracy is a bottleneck." "Eastern Europe Tries to Stoke Up Its Fire Sale," *Business Week*, Oct. 21, 1991, 52.
378. "Hungarian Hotels Sale Delayed by Six Months," *Financial Times*, Oct. 29, 1991; "Fumbled Hotel Sale Snags Hungary's Privatization Scheme," *Reuters Library Report*, Oct. 29, 1991.
379. "Danubius Hotel Chain's Value Revised Downward," *MTI Econews*, Feb. 26, 1992.

380. "Hotel Chain to Come Under the Privatization Hammer," *MTI Econews*, Feb. 13, 1991.
381. "Hungary's Big Sell-Off Hits Hard Times," *Financial Times*, May 21, 1992, I2.
382. "The Painful Road to Privatization," *Financial Times*, Nov. 5, 1991, I29.
383. "Hungary's Privatization Drive in Disarray," Reuters, Oct. 29, 1991.
384. "National Bank of Hungary's Role in Hotel Company Privatization," *MTI Econews*, June 24, 1991.
385. "Danubius Privatization Called Off," *MTI Econews*, Oct. 28, 1991.
386. Ibid.
387. "Fumbled Hotel Sale Snags Hungary's Privatization Scheme," *Reuters Library Report*, Oct. 29, 1991.
388. "Hungarian Hotels Sale Delayed by Six Months," *Financial Times*, Oct. 29, 1991.
389. "Danubius Hotel Chain's Value Revised Downward," *MTI Econews*, Feb. 26, 1992.
390. "Sale of Hotel Chain Postponed After Proposed Change in Tax Break," *BNA International Business Daily*, Nov. 20, 1991.
391. "The Painful Road to Privatisation," *Financial Times*, Nov. 5, 1991, I29.
392. "Marriott Buys Hungarian Hotel Stake," *Financial Times*, July 16, 1992; "Public Placement of Danubius Hotel Shares," *MTI Econews*, July 16, 1992; "Danubius Privatization — Further Details," *MTI Econews*, Sept. 22, 1992.
393. "Public Placement of Danubius Shares," *MTI Econews*, Sept. 3, 1992.
394. See infra Chapter Seven.
395. "Future of SPA Head in Doubt," *MTI Econews*, Aug. 25, 1992.
396. "Privatization in Hungary — an ECONEWS Background," *MTI Econews*, Apr. 25, 1991.
397. "Calling for the Next Stage in Property Rights; An Interview with Dr. Lajos Csepi," *Privat Profit* (Sept. 1991), states: "We strongly suspect that in many cases the transformation has served the interest of those managers who wanted to save themselves and their influence in the future; they did not attempt to make an effort to run their companies in a more effective manner."
398. See supra Chapter Three.
399. State Property Agency, "Information Regarding the Second Privatization Program (Mar. 1991).
400. Ibid.
401. Ibid.
402. Ibid.
403. Data for 1990 in millions of HuF. Source: State Property Agency, "Information Regarding the Second Privatization Program (Mar. 1991).
404. "Calling for the Next Stage in Property Rights: An Interview with Dr. Lajos Csepi," *Privat Profit* (Sept. 1991); "Privatization in Hungary — Konrad Adenauer Foundation," *MTI Econews*, May 27, 1991.
405. Interviews at the State Property Agency, November 1, 1992.
406. Ibid.
407. Some put the number as high as 40,000. See "In Eastern Europe, the Big Sell-Off Is Set to Begin," *Business Week*, Aug. 6, 1990, 42.
408. Emilia Sebok, "Speeding Up Privatization," *Hungarian Economy* 18, nos. 3-4 (1990): 10.
409. Act LXXIV of 1990 on Privatization of State Enterprises Dealing with Detail Trade, Catering Trade, and Consumer Services.

Notes

410. "Hungary Unveils First Phase of Major Privatization Program," *BNA Daily Report for Executives*, Sept. 18, 1990.

411. Indeed, observers were calling the program "a bust" a scant month after it began. See "Hungary's Little Shops of Dreams," *Washington Post*, May 28, 1991, E1, col. 3.

412. "Hungary's Progress Criticised," *East European Markets*, June 26, 1992.

413. "Slow Progress in Hungarian Privatisation," Agence France Presse, Oct. 15, 1991.

414. "Hungary's Progress Criticised," *East European Markets*, June 26, 1992.

415. Nigel Ash, "The Gearbox Needs an Overhaul," *Euromoney Supplement* (April 1992): 41, 47; "Hungary's Progress Criticised," *East European Markets*, June 26, 1992; "Hungary's Little Shops of Dreams," *Washington Post*, May 28, 1991, E1, col. 3.

416. "Hungary's Little Shops of Dreams," *Washington Post*, May 28, 1991, E1, col. 3. Some complained of the SPA's approach as being "heavy handed." See "Slow Progress in Hungarian Privatisation," Agence France Presse, Oct. 15, 1991.

417. Per capita income figures published by the Central Statistical Office in Hungary probably understate the income of many Hungarians badly, because in a cash economy with confiscatory personal income tax rates, much income is simply unreported. Although firm and accurate income figures are impossible to establish, the average Hungarian income is certainly well below that of Western countries.

418. See, for example, "Hungary's Little Shops of Dreams," *Washington Post*, May 28, 1991, E1, col. 3; "Slow Progress in Hungarian Privatisation," Agence France Presse, Oct. 15, 1991; Nigel Ash, "The Gearbox Needs an Overhaul," *Euromoney Supplement* (April 1992): 41; "Hungary's Progress Criticised," *East European Markets*, June 26, 1992.

419. Telling in this regard is the fact that in 1992, Hungarian businesses owed more to Western commercial banks than to Hungarian commercial banks. "Hungary Hankers for Native Capital," *Budapest Week*, Sept. 3-9, 1992, 5, is an interview with SPA consultant Tamas Szabo.

420. Nigel Ash, "The Gearbox Needs an Overhaul," *Euromoney Supplement* (April 1992): 41.

421. For a description of the problem that a lack of credit caused for the auctions, see "Hungary Hankers for Native Capital," *Budapest Week*, Sept. 3-9, 1992, 5.

422. In many of the quaint towns surrounding Lake Balaton, store and restaurant signs are in both German and Hungarian, with the former often given prominence. Houses for rent bear signs "Zimmer Frei" with no Hungarian translation. One Saturday morning in the autumn of 1992, the author and some friends searched one town on the lake for a newspaper in Hungarian but could only find German language papers.

423. "Hungary's Budding Entrepreneurs Bid for Shops and Cafes," *Reuters Library Report*, April 22, 1991.

424. "Hungary's Industry Up for Sale," *Business Law Brief, Financial Times* (Oct. 1990).

425. Ibid.

426. Gerd Schwartz, "Privatization: Possible Lessons from the Hungarian Case," *World Development* 19 (1991): 1731, 1733.

427. "Too Many Firms, Too Few Buyers," *Economist*, Sept. 21, 1991, 14, cites remarks of Dr. Lajos Csepi.

428. State Property Agency, "Information About the Method of Reporting and

Judgement of the Investor-Initiated Privatization of State-Owned Enterprises and State-Owned Shares of Associations (1991).

429. Ibid.
430. Ibid.
431. "Private Possibilities," *Privat Profit* (Sept. 91).
432. State Property Agency, "Information About the Method of Reporting and Judgement of the Investor-Initiated Privatization of State-Owned Enterprises and State-Owned Shares of Associations (1991).
433. Ibid.
434. See infra Chapter Nine.
435. Interviews at the State Property Agency, November 1, 1992.
436. Act VII of 1990, Sec. III.
437. "Private Possibilities," *Privat Profit* (Sept. 91).
438. State Property Agency, "Invitation for Tender on Managing Share and Stake Portfolios Owned by the State Property Agency (April 1991).
439. Ibid.
440. "State Property Agency to Announce New Asset Handling Tender," *MTI Econews*, June 12, 1991.
441. Ibid.
442. "Co-Nexus Active in Managing SPA Portfolio," *MTI Econews*, May 13, 1992.
443. Nigel Ash, "The Gearbox Needs an Overhaul," *Euromoney Supplement* (Apr. 1992): 41.
444. Nigel Ash, "The Gearbox Needs an Overhaul," *Euromoney Supplement* (Apr. 1992): 41–42.
445. "Revlon May Buy Caola," *MTI Econews*, Aug. 13, 1992.
446. "Managing the Managers," *Reason* (March 1991): 48.
447. Ibid.
448. "Eastern Europe Tries to Stoke Up Its Fire Sale," *Business Week*, Oct. 21, 1991, 52.
449. "Proposal on Privatization Strategies in Hungary," *MTI Econews*, June 3, 1991.
450. "Economic Cabinet Discusses Minister of Finance Privatization Strategy," *MTI Econews*, June 18, 1991.
451. "SPA Invites Consultancy Bids for Self-Privatisation Programme," *MTI Econews*, July 8, 1991.
452. "Hungary Expands Streamlined Privatisation Scheme," *Reuters Money Report*, May 5, 1992.
453. "SPA Invites Consultancy Bids for Self-Privatisation Programme," *MTI Econews*, July 8, 1991.
454. "Hungary Expands Streamlined Privatisation Scheme," *Reuters Money Report*, May 5, 1992.
455. "Latest SPA Privatization Tender Attracts 200 Consultancy Firms," *MTI Econews*, July 23, 1991. Applications were submitted by 288 firms, and 265 met the SPA's qualifications. "Self-Privatisation to Start in Hungary Soon," *MTI Econews*, Aug. 6, 1991.
456. Nigel Ash, "The Gearbox Needs an Overhaul," *Euromoney Supplement* (Apr. 1992): 41, 48.
457. "Latest SPA Privatization Tender Attracts 200 Consultancy Firms," *MTI Econews*, July 23, 1991.

458. Ibid.
459. "Self-Privatization to Start in Hungary Soon," *MTI Econews*, Aug. 6, 1991.
460. See "SPA 'Sells' Hungary's Self-Privatization Programme in London," *MTI Econews*, Jan. 24, 1992.
461. "Eastern Europe Tries to Stoke Up Its Fire Sale," *Business Week*, Oct. 21, 1991, 52.
462. See, for example, "Simplified Privatization—A City of London View," *MTI Econews*, May 7, 1992.
463. "SPA "Sells" Hungary's Self-Privatization Programme in London," *MTI Econews*, Jan. 24, 1992.
464. Nigel Ash, "The Gearbox Needs an Overhaul," *Euromoney Supplement* (April 1992): 41, 48.
465. "State Assets Agency Launches Second Phase of Privatisation," *BBC Summary of World Broadcasts*, March 14, 1992.
466. "SPA 'Sells' Hungary's Self-Privatization Programme in London," *MTI Econews*, Jan. 24, 1992.
467. Nigel Ash, "The Gearbox Needs an Overhaul," *Euromoney Supplement* (Apr. 1992): 41, 48.
468. "Hungary Expands Streamlined Privatisation Scheme," *Reuters Money Report*, May 5, 1992.
469. See Gyorgy Matolcsy, "Privatization: Hungary," *Eastern Europe Economics* 30 (Fall 1991): 49, 51–52.
470. "Hungary Eases Sale of State Companies," *Financial Times*, May 6, 1992.
471. "State Assets Agency Launches Second Phase of Privatization," *BBC Summary of World Broadcasts*, Mar. 14, 1992; "Hungary Eases Sale of State Companies," *Financial Times*, May 6, 1992.
472. "Hungary Expands Streamlined Privatisation Scheme," *Reuters Money Report*, May 5, 1992.
473. "Hungary: Privatization Proves to Be a Long, Hard Process," *Inter Press Service*, May 29, 1992.
474. "Hungary: A Race Against Time," *Financial Times*, Oct. 30, 1991.
475. "HAFE to Be Privatized," *MTI Econews*, Feb. 25, 1992.
476. Ibid.
477. "State Liquidation Agency Becomes PLC," *MTI Econews*, Jan. 13, 1992.
478. Ibid.
479. Act IL of 1991 on Bankruptcy and Liquidation Procedures and Final Accounts.
480. "Survey of Privatisation in Eastern Europe," *Financial Times*, July 3, 1992.
481. "New Privatization Program Launched to Break Up 'Hollow' Parent Companies," *BNA International Business Daily*, April 16, 1992.
482. Law LIV of 1992 on the Sale, Utilization, and Protection of Assets Temporarily Owned by the State, sec. 68.
483. "Privatization: Concessions to Domestic Buyers," *MTI Econews*, Dec. 23, 1992.
484. Debevoise and Plimpton, "Central and Eastern European Bulletin, Mar. 23, 1993.
485. Ibid.
486. Ibid.
487. Peter A. Bod, "Deregulation and Institution Building—Lessons from the

Reform of the Hungarian Public Sector," *Jarbuch der Wirtschaft Osteuropas*, band 3, nr. 1 (1990): 120.

488. Peter A. Bod, "Deregulation and Institution Building—Lessons from the Reform of the Hungarian Public Sector," *Jarbuch der Wirtschaft Osteuropas*, band 3, nr. 1 (1990): 120. At the end of 1992, there were 20 commercial banks in Hungary and 10 other financial institutions, including savings banks. "Acquisitions in the East": Europe's New M & A Frontier," *M & A Europe* (Nov.-Dec. 1990).

489. Peter A. Bod, "Deregulation and Institution Building—Lessons from the Reform of the Hungarian Public Sector," *Jarbuch der Wirtschaft Osteuropas*, band 3, nr. 1 (1990): 121. The breakup also gave a boost to the fledgling bond industry by creating a new class of buyers for company bonds and by creating a new class of sellers of bonds. David Bartlett, "The Political Economy of Privatization: Property Reform and Democracy in Hungary," *East European Politics and Societies* 6 (Winter 1992): 73. That expansion of the money supply added inflationary pressure to the economy. Peter A. Bod, "Deregulation and Institution Building—Lessons from the Reform of the Hungarian Public Sector," *Jarbuch der Wirtschaft Osteuropas*, band 3, nr. 1 (1990): 120. That in turn led to a tight-money policy by the HNB, which added to the financial difficulties of the borrowers from the spun-off commercial banks.

490. "Privatization of Large Banks," *Hungarian Observer* (June 1992).

491. "Banking on Privatization: Commercial Banks Are Next in Line to Go Private," *Hungarian Observer* (Oct. 1991).

492. "Hungarian Banks Expected to Be for Sale During Second Half of This Year," *Eastern Europe Report*, Feb. 24, 1992; "Hungarian Banks Could Go on Sale This Year," *Reuters Money Report*, Feb. 5, 1992.

493. Ibid.

494. "The Privitisation Dilemma," *Euromoney* (Sept. 1990): 145–46.

495. "Enthusiasm for Foreign Investment in Hungary Fails to Tell Story," *East European Markets*, July 12, 1991.

496. "Privatization of Large Banks," *Hungarian Observer* (June 1992).

497. "Permanent State Ownership in Commercial Banks," *MTI Econews*, Aug. 31, 1992.

498. Indirect ownership includes ownership by the state-run social security system, ownership by state owned firms of which the state owns a majority, and ownership by state-sponsored cooperatives. "Permanant State Ownership in Commercial Banks," *MTI Econonews*, Aug. 31, 1992.

499. "Hungarian Banks Could Go on Sale This Year," *Reuters Money Report*, Feb. 5, 1992.

500. "Hungary: Privatization Proves to Be a Long, Hard Process," Inter Press Service, May 29, 1992.

501. "Hungary: A Giant Step Ahead," *Business Week*, Apr. 15, 1991, 58.

502. "Hungary Hankers for Native Capital," *Budapest Week*, Sept. 3-9, 1992, 5. In fairness to the banks, there was a paucity of credit-worthy businesses to which to lend in Hungary in the late 1980s and early 1990s. In addition, high inflation and forint devaluations forced interest rates paid on deposits to remain high during the period, making it difficult for the banks to make attractively priced loans to businesses. See *Budapest Week*, Oct. 8-14, 1992.

503. Speech of Sir William Ryrie in Czechoslovakia (July 1991).

504. "Compare "Credit Consolidation Programme Approved," *MTI Econews*, Dec. 18, 1992, which quotes $1.6 billion figure, with "Hungary Readies Bank Bailout Scheme," *Reuters Money Report*, Dec. 18, 1992, which quotes $3 billion figure.

Notes

505. "Report of the State Banking Supervision," *MTI Econews*, Oct. 30, 1992.
506. Ibid.
507. The aggregate numbers were pulled down by the collapsing state sector and did not reflect the vitality of the Hungarian economy in late 1992. The private entrepreneurial sector, which by some estimates accounted for as much as 50 percent of Hungary's GDP in the second half of 1992, was growing at an estimated rate of 25 percent per year. See William Echikson, "Where Eastern Europe Is Booming," *Fortune*, July 12, 1993.
508. "Privatization of Hungarian Commercial Banks," *MTI Econews*, Jan. 14, 1992.
509. "Privatization of Large Banks," *Hungarian Observer* (June 1992).
510. "Prodded by New Law, Hungarian Banks May Edge Away from State Control Toward Privatization," *EBRD Watch* 2, no. 32 (Aug. 31, 1992): 3.
511. Act LXIX of 1991 on Banks and Banking Activities, arts. 18(1), 97(3), discussed in "Privatization of Hungarian Banks: State Must Reduce Ownership Share by Law," *MTI Econews*, Nov. 26, 1991; "Privatization of Large Banks," *Hungarian Observer* (June 1992).
512. Act LXIX of 1991 on Banks and Banking Activities, art. 97(4), discussed in "Permanent State Ownership in Commercial Banks," *MTI Econews*, Aug. 31, 1992.
513. Law LXIX of 1991 on Banks and Banking Activities, arts. 21-23, 98-101, discussed in "Privatization of Hungarian Commercial Banks," *MTI Econews*, Jan. 14, 1992.
514. Act LX of 1991 on the National Bank of Hungary, art. 84(1), discussed in "Prodded by New Law, Hungarian Banks Edge Away from State Control Toward Privatization," *EBRD Watch*, 2, no. 32, p. 3 (Aug. 31, 1992): 3.
515. "Legislation on Privatization and Acquisitions," *Doing Business with Eastern Europe*, Aug. 1, 1991.
516. "Hungary Lists Firms That Will Not Be Privatized," *Reuters Money Report*, May 28, 1991.
517. Act LX of 1991 on the National Bank of Hungary, art. 52(2).
518. "Prodded by New Law, Hungarian Banks Edge Away from State Control Toward Privatization," *EBRD Watch*, 2, no. 32 (Aug. 31, 1992): 3.
519. Law LXIX of 1991 on Banks and Banking Activities; Act LX of 1991 on the National Bank of Hungary.
520. "Hungarian Banks Could Go on Sale This Year," *Reuters Money Report*, Feb. 5, 1992.
521. "Bank Privatization Committee Established," *MTI Econews*, April 6, 1992.
522. "Prodded by New Law, Hungarian Banks Edge Away from State Control Toward Privatization," *EBRD Watch*, 2, no. 32 (Aug. 31, 1992): 3.
523. Ibid.
524. "New Directorate and Supervisory Committee Appointed for the Credit Bank," *BBC Summary of World Broadcasts*, May 8, 1992.
525. "Privatization of Large Banks," *Hungarian Observer* (June 1992).
526. "EBRD Chasing Stakes in Hungary's Commercial Banks," *Reuters Money Report*, Sept. 29, 1992.
527. Ibid.
528. "Bank Privatization Adviser to Be Chosen in October," *MTI Econews*, Sept. 8, 1992.

529. Ibid.
530. "SPA Official on Selection of Bank Privatization Adviser," *MTI Econews*, Sept. 8, 1992.
531. "Shortlist of Bank Privatization Consultants Published," *MTI Econews*, Sept. 30, 1992.
532. "Business Briefs," *Daily News* (Budapest), Nov. 6–12, 1992, 4.
533. "Banking on Privatization: Commercial Banks Are Next in Line to Go Private," *Hungarian Observer* (Oct. 1991).
534. "State Acquires Majority in Konzumbank," *MTI Econews*, Dec. 12, 1992.
535. "The Domestic Money and Stock-Market," *Business and Economy/Invest in Hungary* (EGISZ Nov. 1992).
536. "Bank Inspectorate Rules Against Ybl Bank," *MTI Econews*, July 2, 1992.
537. "Ybl Bank's Securities Trading Rights Suspended," *MTI Econews*, July 7, 1992.
538. Ibid.
539. "Three Hungarian Banks Face Bankruptcy," *MTI Econews*, July 8, 1992.
540. "BSE Board Suspends Ybl Bank's Membership," *MTI Econews*, July 17, 1992.
541. "Will Bankrupt Banks Survive? Supervision Gets Proposals," *MTI Econews*, July 27, 1992.
542. "Three Banks Show Interest in Ybl Bank," *MTI Econews*, Dec. 9, 1992.
543. Ibid.
544. "Budapest Police and Prosecutor's Office Data on Corruption in the Banking Sector," *BBC Hungarian Radio,* Sept. 14, 1993. See also "Police Investigating Banking Fraud to Value of HuF 20 Bn.," *MTI Econews*, Sept. 9, 1993, which reports alleged misuse of approximately 6 billion HuF in depositors' funds by Ybl.
545. Ibid.
546. "The Domestic Money and Stock-Market," *Business and Economy/Invest in Hungary* (EGISZ Nov. 1992).
547. "Depositors Sue State Banking Supervision," *MTI Econews*, Mar. 12, 1993.
548. "The Domestic Money and Stock-Market," *Business and Economy/Invest in Hungary* (EGISZ Nov. 1992).
549. Ibid.
550. "Three Hungarian Banks Face Bankruptcy," *MTI Econews*, July 8, 1992.
551. "The Domestic Money and Stock-Market," *Business and Economy/Invest in Hungary* (EGISZ Nov. 1992).
552. Ibid.
553. "Will Bankrupt Banks Survive? Supervision Gets Proposals," *MTI Econews*, July 27, 1993.
554. "Budapest Police and Prosecutor's Office Data on Corruption in the Banking Sector," *BBC Summary of World Broadcasts*, Aug. 14, 1993.
555. "Hungary Probes Bank Fraud," *Wall Street Journal*, Sept. 10, 1993, A3, col. 4.
556. Ibid.
557. "42 Cases of Bank Fraud Under Investigation," *MTI Econews*, Sept. 13, 1993.
558. "Hungarian Police Investigate Banking Scandal," UPI, Sept. 9, 1993.
559. "Interview with Erich Wehr," March 1993. Wehr was at that time director of Westdeutschelandesbank (Hungary) Rt., a subsidiary of Westdeutschelandesbank

that was a 59 percent owner of AVB. As such, he was in control of AVB on behalf of the parent company.

560. Ibid.
561. Ibid.
562. Ibid.
563. Ibid.
564. See "Hungary Readies Bank Bail-Out Scheme," *Reuters Money Report*, Dec. 18, 1992; "Credit Consolidation Programme Approved," *MTI Econews*, Dec. 18, 1992.
565. Source: Hungarian Central Statistical Office.
566. Wayne Ringlien, "Privatizing of Farming Activities in Hungary," *Finance and Development* (Dec. 1990).
567. Ibid.
568. "Hungary Plows New Territory: State Farms Follow a Bumpy Road to Privatization," *Chicago Tribune*, Mar. 20, 1992.
569. "State Farms to Be Privatized," *MTI Econews*, July 23, 1992.
570. "Hungary Plows New Territory: State Farms Follow a Bumpy Road to Privatization," *Chicago Tribune*, Mar. 20, 1992.
571. These figures illustrate an inefficiency that plagues the Hungarian economy generally: gross overstaffing. On the Monor farm, typical of state farms in Hungary, there was one worker for every 4.4 acres. By comparison, in the United States, approximately 9.6 million farm employees work approximately 991 million acres, an average of one worker for every 103 acres. "Bureau of Labor Statistics," U.S. Department of Labor (1990).
572. Ibid.
573. Bureau of Labor Statistics, U.S. Department of Labor (1990). The general tendency of Hungarian businesses to be conglomerated has made it difficult to sell them as going concerns because Western businesses, which once saw conglomeration as a method of eliminating unsystematic risk, now look with disfavor upon conglomeration as unwieldy and inefficient. See "Hungary: Privatization Proves to Be a Long, Hard Process," Inter Press Service, May 29, 1992.
574. "Hungary Lists Firms That Will Not Be Privatized," *Reuters Money Report*, May 28, 1991.
575. "Proposals on Privatization Strategies in Hungary," *MTI Econews*, June 3, 1991.
576. "BSE to Establish Clearing House," *Daily News* (Budapest), Oct. 2-8, 1992, 4.
577. This problem has always been more one of perception than of reality. In mid-1991, foreigners owned only approximately 3 to 4 percent of Hungarian industry compared with 25 to 30 percent in other European countries such as Belgium and Austria and similar percentages in pre-World War II Hungary. See "Calling for the Next Stage in Property Rights; An Interview with Dr. Lajos Csepi," *Privat Profit* (Sept. 1991).
578. "Hungarians Need Bigger Role in Privatisation: SPA," *Reuters Money Report*, Jan. 13, 1992.
579. The government of Hungary had long said that it wanted to privatize more than 50 percent of state assets by 1994. See, for example, "State to Keep Ownership of 100 Enterprises," *MTI Econews*, Oct. 7, 1992.
580. Ibid.
581. See, for example, O. Blanchard and R. Layard, "Economic Change in

Poland," Discussion Paper No. 424, Center for Economic Policy Research, London, 1990; David M. Newberry, "Reform in Hungary—Sequencing and Privatisation," *European Economic Review* 35 (1991): 571, 578-79; "Hungary's Progress Criticised," *East European Markets*, June 26, 1992, which cites a study by the Privatisation Research Institute.

582. See, for example, "Sachs, The Property Distributor," *Privat Profit* (Aug. 1991).

583. See, for example, "Hungary's Progress Criticised," *East European Markets* (June 26, 1992) (citing study by Privatisation Research Institute).

584. "Unhappy Hunting Ground," *International Management* (Sept. 1990): 63.

585. See Stark, "Privatization in Hungary: From Plan to Market or from Plan to Clan?" *East European Politics and Societies* 4 (Fall 1990): 351, 370-71.

586. "Steady Growth in Personal Savings," *MTI Econews*, Sept. 17, 1992.

587. "Hungary Hunkers for Native Capital," *Budapest Week*, Sept. 3-9, 1992, 5.

588. Ibid.

589. Interview with Dr. Zsuzsanna Zatik, October 1992.

590. "Not Selling Out, But Looking for Partners," EGISZ (July-Sept. 1989).

591. "Proposed Privatization Strategies in Hungary," *MTI Econews*, June 3, 1991, outlines government draft policy on privatization, including a platform calling for distribution of property for free only in very limited special circumstances.

592. See, for example, State Property Agency, "Privatization and Foreign Investment in Hungary" (Mar. 1991): 6. SPA officials remained adamant through 1992 about the government's policy in favor of sales. Interview with Dr. Jeno Czuczai, State Property Agency (Nov. 1992).

593. See, for example, Suzanne Loeffelholz, "Dealing on the Danube," *Financial World,* Mar. 6, 1990, which quotes government economist Laszlo Antall as saying Hungary wanted to avoid the radical and chaotic Polish example.

594. "Hungary Makes Striking Switch to Mass Privatisation," *Financial Times*, Oct. 6, 1992.

595. "The Privatizing of Eastern Europe," *Institutional Investor* (April 1990):175.

596. Ibid.

597. "Hungary Unveils First Phase of Major Privatization Program," *BNA Daily Report for Executives* (Sept. 18, 1990).

598. "Free Market Changes Underway," *Hungary Handbook* (Nov. 1990).

599. "Privatization Loan in Hungary," *MTI Econews*, Feb. 8, 1991.

600. Ibid.

601. Ibid.

602. "Existence Credit Becomes Cheaper," *MTI Econews*, Jan. 24, 1992.

603. See, for example, Nigel Ash, "The Gearbox Needs an Overhaul," *Euromoney Supplement* (April 1992): 41.

604. "Spokesman's Briefing—Loans for Privatisation," *MTI Econews*, Jan. 24, 1992. According to one source, the two funds combined for loans less than 1 billion HuF ($13.3 million) in 1991. *BBC Summary of World Broadcasts*, Jan. 18, 1992.

605. "Spokesman's Briefing—Loans for Privatisation," *MTI Econews*, Jan. 24, 1992.

606. "Existence Credit Becomes Cheaper," *MTI Econews*, Jan. 24, 1992.

607. Ibid.

608. "Privatization Loan Credits Modified," *MTI Econews*, April 9, 1992.

609. "Privatization to Gain Momentum Next Year, Says Minister," *MTI Econews*, Dec. 22, 1992. The minister of finance also began negotiations with the

Hungarian National Bank to reduce the interest rate to 3 percent, with the difference between that figure and the National Bank's cost of funds to be paid for out of the state budget. See "Reduced E-Credit Interest," *MTI Econews*, Dec. 11, 1992.

610. Ibid.

611. "Existence Credit Becomes Cheaper," *MTI Econews*, Jan. 24, 1992. Indeed, E-Credits were used not only to start businesses but to finance the purchase of owner-occupied residential real estate. Interview with Antonia Szabo, February 12, 1993; interview with Karoly Bognar, March 6, 1993.

612. Interview with Karoly Bognar, March 6, 1993. Interview with Antonia Szabo, October 1992. Szabo is a government official who in 1992 opened a small boutique in the fashionable Vaci utca with the help of an E-Credit loan.

613. Interview with Dr. Karoly Bognar, September 1992.

614. "Hungary Readies Privatisation 'Credit Cards,'" *Reuters Library Report*, Oct. 12, 1992.

615. "Privatization Loan Coupon," *MTI Econews*, Oct. 14, 1992.

616. "Private Possibilities," *Privat Profit* (Sept. 1991). They were also eligible, like any other domestic investors, for E-Credits, which gave them a double advantage.

617. "Privatization: Concessions to Domestic Buyers," *MTI Econews*, Dec. 23, 1992.

618. The SPA planned to retain 20 percent of the shares to make them available for compensation coupons. "Employee and Management to Buyout Kanizsa Brewery," *MTI Econews*, July 24, 1992.

619. "Employee and Management to Buyout Kanizsa Brewery, *MTI Econews*, July 24, 1992. Another $2.7 million was to come from an E-Credit; the rest of the consideration was undisclosed.

620. "AGRROINVEST Employees May Buy Shares," *MTI Econews*, Nov. 30, 1992.

621. Ibid.

622. Act XLIV of 1992 on the Employees' Part-Ownership Program, preamble.

623. Ibid., sec. 1.

624. Ibid., sec. 2.

625. Ibid., secs. 2, 15.

626. Ibid., sec. 9.

627. Ibid., sec. 25.

628. See, for example, "Parliament Passes ESOP Law," *MTI Econews*, June 10, 1992.

629. "Government Gives Boost to Home Grown Capitalism," *Financial Times*, May 29, 1992.

630. Ibid.

631. Ibid.

632. "Hungarian proposal Would Favor Domestic Over Foreign Investors," *BNA International Financial Daily*, June 4, 1992.

633. Ibid.

634. These privatizations were also significant in that they saw the reemergent use of the Budapest Stock Exchange as a medium of privatization sales. After the IBUSZ fiasco, discussed above, and the weakness of prices and thinness of float on the BSE, the SPA had largely ignored the BSE, much to the chagrin of stock exchange officials, whose struggling exchange needed the business and prestige of public offerings.

635. "Public Issue of Pick Salami Shares to Start October 27," *MTI Econews*, Sept. 30, 1992.
636. "Marriott Buys Hungarian Hotel Stake," *Financial Times*, July 16, 1992.
637. "Pick Salami Shares Aimed at Small Investors," *MTI Econews*, July 16, 1992.
638. Ibid.
639. "Public Issue of Pick Salami Shares to Start October 27," *MTI Econews*, Sept. 30, 1992. In practice, there was no apparent enforcement of the provision that the individual be a Hungarian citizen. One Austrian claimed to have purchased shares on a preferential basis from an established Budapest brokerage firm on the morning of the public offer. Because the Austrian was well-known in the financial community, there was no question that the brokerage firm knew of his foreigner status. Interview with the managing director, management consulting firm, November 2, 1992. In fact, the desire to sell the stock seemed to consume those handling the sales. Glossy color posters touted the stock in the post offices. And on the morning the stock went on sale, a representative of the lead underwriter was interviewed on a local radio station. The representative told the interviewer that there were long lines at all the Budapest outlets where the stock was for sale but that the lines outside Budapest were shorter. There was, he noted, tremendous demand for the stock. In fact, personal inspection showed no lines at any of the outlets in Budapest. Indeed, the stock was available with no problem or waiting several days after it first went on sale. The underwriter seems to have been attempting to create a stampede.
640. See supra pages 31-32.
641. "Public Placement of Danubius Hotel Shares," *MTI Econews*, July 16, 1992.
642. "Last Breath of the Pickled Cucumber," *Times* (London), Nov. 21, 1991.
643. "Danubius Privatization—Further Details," *MTI Econews*, Sept. 22, 1992; "Public Placement of Danubius Hotel Shares," *MTI Econews*, July 16, 1992; "Marriott Buys Hungarian Hotel Stake," *Financial Times*, July 16, 1992; "Hungary Hotel Sale to Encourage Local Investors," *Reuters Business Report*, July 15, 1992.
644. "Hungarians Need Bigger Role in Privatisation—SPA," *Reuters Money Report*, Jan. 13, 1992.
645. "Hungary: Uses of Compensation Coupons," *BBC Summary of World Broadcasts*, May 7, 1992.
646. "Hungary Acts to Speed Up Flagging Privatisation," *The Reuter Library Report*, Dec. 11, 1992.
647. "Hungary Makes Striking Switch to Mass Privatisation," *Financial Times*, Oct. 9, 1992.
648. "Hungary Hankers for Native Capital," *Budapest Week*, Sept. 3-9, 1992, cites 85 percent figure; Nigel Ash, "The Gearbox Needs an Overhaul," *Euromoney Supplement* (April, 1992): 41, cites 90 percent figure.
649. "Hungary Hankers for Native Capital," *Budapest Week*, Sept. 3-9, 1992, 5.
650. Ibid.
651. Interview with Dr. Jeno Czuczai, November 1992.
652. In 1992 twice as many foreign firms were involved in some way in the process of privatization as in 1990 and 1991 combined. "Privatization to Gain Momentum Next Year, Says Minister," *MTI Econews*, Dec. 22, 1992.
653. A poll released in early 1993 showed the MDF behind the opposition FIDESZ party as Hungary prepared for elections in 1994. See *Budapest Post*, April 22, 1993, 1.

654. "Privatization to Gain Momentum Next Year, Says Minister," *MTI Econews*, Dec. 22, 1992. It is important to remember that just because a company has been "privatized" does not mean that the state has wholly divested its interest. Indeed in most cases the SPA retained an interest in privatized firms. In addition, shares were often reserved for the state social security fund or reserved for later exchange for compensation coupons or sale to employees at a reduced price.

655. "Privatization: Concessions to Domestic Buyers," *MTI Econews*, Dec. 23, 1992.

656. "Hungary Steps Up Sell-Offs," *MTI Econews*, Dec. 23, 1992.

657. "Economic Affairs in Brief," *BBC Summary of World Broadcasts*, Jan 31, 1992.

658. Ibid.

659. "Privatization: Concessions to Domestic Buyers," *MTI Econews*, Dec. 23, 1992. A larger number of former SOEs, however, had been at least partially privatized or liquidated. According to an SPA spokesperson speaking in late 1992, there were approximately 1,660 SOEs remaining that had not been through privatization at all. "Privatization Slows in Hungary," *MTI Econews*, Aug. 13, 1992. If true, approximately 30 percent of the roughly 2,400 SOEs had been partially privatized or liquidated in the period 1990-1992.

660. "SPA Report On 1991," *MTI Econews*, July 9, 1992.

661. See "Flood of Foreign Investment Capitalizes on New Hungary," *Washington Post*, Nov. 10, 1991, H1.

662. "Compare State Property Agency, "Information Booklet (1991), which reports a $26.7 billion figure, with "On the Back Burner," *Hungarian Observer* (Oct. 1990), which reports a $37 billion figure.

663. The figures break down as follows: **1990:** $8 million. See "Hungary Unable to Collect Target Revenue Through Privatization," Xinhua General Overseas News Service, Sept. 10, 1991. **1991:** $418 million. See "Hungary: Income From Privatisation in First Five Months of 1992," *BBC Summary of World Broadcasts*, July 2, 1992. **1992:** $780 million. See "Privatization: Concessions to Domestic Buyers," *MTI Econews*, Dec. 23, 1992.

664. "Government Approves Plan to Speed Up Privatization," *MTI Econews*, Dec. 11, 1992.

665. Ibid.

666. "Hungary's Progress Criticised," *East European Markets*, June 26, 1992.

667. "Flood of Foreign Investment Capitalizes on New Hungary," *Washington Post*, Nov. 10, 1991, H1.

668. "Survey of Privatisation in Eastern Europe," *Financial Times*, July 3, 1992.

669. "Finance Minister Says Privatisation at Standstill for Last 6 Months," *BBC Summary of World Broadcasts*, Oct. 6, 1992.

670. Ibid.

671. "Government on Economic Issues; SPA on Privatization Speed-Up," *MTI Econews*, Dec. 11, 1992.

672. Nigel Ash, "The Gearbox Needs an Overhaul," *Euromoney Supplement* (Apr. 1992): 41, 49.

673. "World Economy 6; Focus on Privatisation in Eastern Europe," *Financial Times*, Oct. 14, 1991.

674. Ibid.

675. "Flood of Foreign Investment Capitalizes on New Hungary," *Washington Post*, Nov. 10, 1991, H1.

676. "Hungary Now Poor Third in Investment Race," *Budapest Sun*, Nov. 25–Dec. 1, 1993, 1. The cited article was the second in a two-part series entitled "Privatization Stalled."

677. "Privatization: Concessions to Domestic Buyers," *MTI Econews*, Dec. 23, 1992. Curiously, at the same time these figures were being released, SPA officials were continuing to suggest that privatization would "accelerate in 1993. See "Privatization to Gain Momentum Next Year, Says Minister," *MTI Econews*, Dec. 22, 1992.

678. "New Privatization Forms," *Daily News* (Budapest), Nov. 13–19, 1992, 4.

679. Act LIII of 1992 on the Management and Utilization of Entrepreneurial Assets Permanently Remaining in State Ownership.

680. "Hungarian Firms Seek Partners," *Eastern European Report*, Dec. 23, 1991.

681. "Alitalia Acquires 30 Pct. of Hungarian Airline Malev," *AFX News*, Dec. 15, 1992.

682. "Privatization: Concessions to Domestic Buyers," *MTI Econews*, Dec. 23, 1992.

683. "Government on Economic Issues; SPA on Privatization Speed-Up," *MTI Econews*, Dec. 11, 1992.

684. "Hungary Steps Up Sell Offs," *MTI Econews*, Dec. 23, 1992.

685. Ibid.

686. "Hungary: Need for Western Capital," *Financial Times*, Oct. 30, 1991, quotes Peter Zelnick of Girozentrale.

687. The story of the Budapest Commodity and Capital Exchange is told in detail in Katalin Mero, *The Role of the Stock Market and Its Significance in the Economic Life of Capitalist Hungary (1864–1944)*, trans. Zsuzsanna Magulya. In addition, Bela Jancso, who was a broker at the Budapest Commodity and Capital Exchange in the early 1940s, provided valuable information in an interview conducted in February 1993.

688. Buda and Pest were still separate cities at the time. They were officially merged into Budapest in 1872. Nonetheless, the names of organizations operating in the metropolitan area reflected the commercial reality of their unity much earlier.

689. Pursuant to legislation adopted in that year, the Budapest Chamber of Commerce authorized a limited number of qualified persons to act as financial intermediaries in commercial transactions. These persons—the sworn brokers—could not act as principals on their own behalf and were sworn to secrecy regarding transactions they facilitated. When the Budapest Commodity and Capital Exchange opened in 1864, only sworn brokers were permitted to execute transactions there. The institution was not abolished officially until 1875, although in practice its exclusivity was substantially eroded almost immediately by the failure of anyone to enforce the rule. Exchange rules promulgated in 1869 officially permitted trading by other "licensed brokers"—a new category—and in 1875 the Hungarian Parliament officially did away with the sworn broker concept.

690. That building, which sits just across Szabadsag Square from the American Embassy, now houses Magyar Televizio (MTV), the Hungarian state television station.

691. At the same time the stock market was slumping, the value of government bonds was shooting up, more than doubling in value from 1883 to 1894.

692. David Bartlett, "The Political Economy of Privatization: Property Reform and Democracy in Hungary," *East European Politics and Societies* 6 (Winter 1992): 73, 101.

693. Ibid.
694. Bradley Graham, "Flirting with Capitalism; Experiment in Hungary," *Washington Post*, Oct. 19, 1983, A1.
695. Peter A. Bod, "Deregulation and Institution-Building Lessons from the Reform of the Hungarian Public Sector," *Jarbuch der Wirtschaft Osteuropas*, band 13, nr. 1 (1990). See "Disintermediation, Hungarian Style," *Forbes*, July 4, 1983, 77.
696. "Hungary's Savvy Banker," *Time*, June 27, 1983, 50.
697. *Reuters North European Service*, May 9, 1983.
698. "Hungary Opens Bond Market for State Enterprises," *Financial Times*, Mar. 30, 1983, I1; David Bartlett, "The Political Economy of Privatization: Property Reform and Democracy in Hungary," *East European Politics and Societies* 6 (Winter, 1992): 73, 101; "Bull's Blood Bonds," *Economist*, Apr. 16, 1983, 89; "Disintermediation, Hungarian Style," *Forbes*, July 4, 1983, 77.
699. David Buchan, "Phone Offer to Hungarian Bond Subscribers," *Financial Times*, Dec. 7, 1983, I3. Other bonds issued by the Post Office at about the same time and without the telephone inducement carried an 11.5 percent coupon.
700. "Local Telephone and School Bonds," *BBC Summary of World Broadcasts*, Jan. 16, 1986.
701. "Hungarian Bonds," *The Economist*, Mar. 15, 1985, 86.
702. Michael T. Kaufman, "Hungarians Clip Coupons and Call It Communism," *New York Times*, July 8, 1986, A2, col. 3.
703. Jonathan Lynn, "International News," Reuters, Sept. 2, 1985.
704. Ibid.
705. Ibid.
706. David Buchan, "Hungary's Bond Trading Obtains the Ring of Success," *Financial Times*, Jan. 24, 1986, I122.
707. "Local Telephone and School Bonds," *BBC Summary of World Broadcasts*, Jan. 16, 1986.
708. Michael T. Kaufman, "Hungarians Clip Coupons and Call It Communism, *New York Times*, July 8, 1986, A2, col. 3.
709. As described in Chapter Six, in 1987 the commercial and savings bank functions of the Hungarian National Bank were spun off into several separate private commercial and savings banks.
710. "Hungarian Banks Agree Securities Market Framework," *Reuters Money Report*, Jan. 12, 1988.
711. "Hungarian Banks Agree to Regulate Domestic Securities Market," *The Reuters Library Report*, Jan. 12, 1988.
712. Ibid.
713. "Hungary; Stock Exchange Council Elected," *BBC Summary of World Broadcasts*, Aug. 4, 1988.
714. "Government Supports Securities Market," *Business and Economy/Invest in Hungary*, July-Sept. 1989.
715. "Hungary; Stock Exchange Council Elected," *BBC Summary of World Broadcasts*, Aug. 4, 1988.
716. "Annex to the Act on Securities and the Stock Exchange," *Public Finance in Hungary* 64 (1990): 106, 108.
717. Leslie Colitt, "Hungary Reopens the Doors of Its Closed Economy," *Financial Times*, April 4, 1989, I3.
718. "Government Supports Securities Market," *Business and Economy/Invest in Hungary* (July-Sept. 1989).

719. Ibid.

720. Act VI of 1990 on Securities and the Stock Exchange regulated the initial distribution of securities, the broker-dealer industry, and secondary trading markets. It also authorized the establishment of securities exchanges that could meet certain criteria.

721. In September 1990, Zsigmond Jarai left the Finance Ministry to begin a lucrative career in the securities brokerage business.

722. "Annex to the Act on Securities and the Stock Exchange," *Public Finance in Hungary* 64 (1990): 106-7.

723. Steven Greenhouse, "Lonely Days for Traders at Budapest Exchange," *New York Times*, Feb. 20, 1990, D1, col. 3.

724. "Annex to the Act on Securities and the Stock Exchange," *Public Finance in Hungary* 64 (1990): 106, 110.

725. Steven Greenhouse, "Lonely Days for Traders at Budapest Exchange," *New York Times*, Feb. 20, 1990, D1, col. 3.

726. "Annex to the Act on Securities and the Stock Exchange," *Public Finance in Hungary* 64 (1990): 106.

727. Ibid.

728. Ibid.

729. "The First Two Years of the Operation of the Budapest Stock Exchange (June 21, 1990–June 19, 1992)," Budapest Stock Exchange (1992).

730. Construction delays for even the smallest projects were legendary in Budapest.

731. Interview with Lyman Johnson, January 1991.

732. "The First Two Years of the Operation of the Budapest Stock Exchange (June 21, 1990–June 19, 1992)," Budapest Stock Exchange (1992).

733. Act VI of 1990 on Securities and the Stock Exchange, secs. 42-74.

734. *Charter of the Budapest Stock Exchange*, June 19, 1990, 91-92. Two banks not present at the founders' meeting on June 19 signed the charter at the opening ceremony on June 21.

735. *Charter of the Budapest Stock Exchange*, June 19, 1990, 24.

736. Ibid.

737. Ibid., 27, 31-40.

738. Ibid., 41-42.

739. Ibid., 47-49.

740. Ibid., 44-45.

741. Ibid., 66.

742. See, for example, Interview with Dr. Kalman Debreceni, January 1991. Debreceni was the head of a small investment banking firm and a founding member of the Stock Exchange Council.

743. Interview with Tamas Lovas, January 1991.

744. Interview with Viktoria Biro, November 1992.

745. Interviews with Tamas Lovas and Mariann Vida of the Budapest Stock Exchange, January 1991.

746. *Rules of the Budapest Stock Exchange Regarding the Requirements of the Listing and Trading of Securities on the Stock Exchange* (1990).

747. Ibid.

748. Interview with Tamas Lovas, January 1991; interview with Marianne Vida, January 1991.

749. *Budapest Stock Exchange Factbook, 1991.*

750. Established in 1989, Dunaholding began as a traded share on the BSE on November 19, 1990, and changed to the listed category on September 16, 1991. It was established as a holding company with major investments in eight different ventures. *Budapest Stock Exchange Factbook, 1991.*

751. Fotex was established in 1984, became a company limited by shares (Rt.) in 1990, was traded on the BSE as of November 13, 1990, and was listed as of March 1, 1991. It was originally founded as a series of photo-finishing stores, but it diversified into a conglomerate. *Budapest Stock Exchange Factbook, 1991.*

752. IBUSZ was the official state travel agency for Hungary. It was founded in 1902 and was the first share with a BSE listing. It started on the exchange as a listed share and was the first to trade on June 21, 1990, when the exchange opened.

753. Konzum, a holding company with various investments, began as a listed share on November 1, 1990.

754. Styl, a manufacturer of ready-to-wear clothing, was founded in 1989 and became a listed share on the day of the Budapest Stock Exchange's opening, June 21, 1990. *Budapest Stock Exchange Factbook, 1991.*

755. Sztrada Skala, a holding company with a majority stake in Skala Coop, a department store chain, was formed in 1988 and became listed on the BSE on January 7, 1991.

756. Zalakeramia, a ceramics manufacturer, was founded in 1991 and on August 1 of that year became a listed share on the BSE. *Budapest Stock Exchange Factbook,* 1991

757. Founded in 1989, Agrimpex, an import-export firm, began trading on the BSE on July 1, 1991.

758. Bonbon Hemingway was founded in 1990 as a confectioner. It began trading on the Budapest Stock Exchange on the day of the exchange's opening, June 21, 1990.

759. Budaflax, a textile manufacturer, was founded in 1988 and became a listed company on June 11, 1991. *Budapest Stock Exchange Factbook, 1991.*

760. This company, with its rather cumbersome name, was a brewery company founded in 1988. Its shares began trading on the BSE on October 29, 1990. *Budapest Stock Exchange Factbook, 1991.*

761. Another diversified trading company, Garagent was founded in 1988, and its shares began trading on the BSE on December 19, 1991.

762. Hungagent, a diversified holding company, was founded in 1968 and began to be traded on the BSE on December 5, 1991. *Budapest Stock Exchange Factbook, 1991.*

763. Founded in 1991, Kontrax Irodatechnika was established as a business supply firm. Its shares began trading on the BSE in 1991.

764. Another business supply trading firm, Kontrax Telecom was founded in 1991, and its shares began trading on the BSE the same year. *Budapest Stock Exchange Factbook, 1991.*

765. Muszi, a holding company with interests primarily in the agricultural sector, was established in 1969, and its share began trading on the Budapest Stock Exchange in 1991.

766. Nitroil, a chemical manufacturer, was founded in 1989, and its shares began trading on the BSE in February 1991. *Budapest Stock Exchange Factbook, 1991.*

767. Novotrade was formed in 1984 as a holding company for a variety of trading firms. Its shares began trading on the BSE on April 9, 1991. *Budapest Stock Exchange Factbook, 1991.*

768. Skala-Coop, formed in 1988, was primarily the operator of a chain of department stores in Budapest. Its shares began trading on the Budapest Stock Exchange in January 1991. *Budapest Stock Exchange Factbook, 1991.*

769. Terraholding was a holding company formed in 1990, with subsidiaries operating in a number of areas of trade. Its shares began trading on the BSE on July 11, 1991. *Budapest Stock Exchange Factbook, 1991.*

770. *Budapest Stock Exchange Factbook, 1992–June, 1993.*

771. Danubius Hotel Company became a listed company on the BSE on December 23, 1992. *Budapest Stock Exchange Factbook, 1991.*

772. Fonix, a diversified trading company, was founded in 1989, and its shares began trading on the BSE on January 13, 1992.

773. Pick Szeged, established in 1869 and transformed into a company limited by shares in June 1992, was primarily a salami manufacturer. It was traded on the BSE as of December 21, 1992, following a public offering of its shares. *Budapest Stock Exchange Factbook, 1992–June, 1993.*

774. *Budapesti Ertektozsde 1991.*

775. *Rules of the Budapest Stock Exchange Regarding the Types of Transactions That May be Carried Out on the Stock Exchange and Trading on the Floor of the Stock Exchange* (1990).

776. Ibid.

777. Ibid.

778. Law LXIX of 1991 on Investment Funds.

779. "Plans for Compensation Certificate-for-Share Exchange," *MTI Econews*, Dec. 9, 1991.

780. "Financial Futures Market," *Budapest Daily News*, Nov. 13–19, 1992, 4; "Financial Futures Market to Start in March," *MTI Econews*, Nov. 18, 1992.

781. "Regulations of the Settlement Department of the Budapest Stock Exchange" (1990).

782. Interview with Viktoria Biro, November 1992.

783. *Budapest Week*, Oct. 8–14, 1992, 4.

784. *MTI Econews.*

785. *MTI Econews.*

786. *The Three Years of the Budapest Stock Exchange, June 21, 1990–May 31, 1993*, Budapest Stock Exchange (1993).

787. Stark, "Privatization in Hungary: From Plan to Market or from Plan to Clan?" *East European Politics and Societies* 4 (Fall 1990): 351, 369.

788. David Bartlett, "The Political Economy of Privatization: Property Reform and Democracy in Hungary," *East European Politics and Societies* 6 (Winter 1992): 73, 106.

789. "Slow Progress in Hungarian Privatisation," Agence France Presse, Oct. 15, 1991.

790. David Bartlett, "The Political Economy of Privatization: Property Reform and Democracy in Hungary," *East European Politics and Societies* 6 (Winter 1992): 73, 106.

791. "On the Back Burner," *Hungarian Observer* (Oct. 1990).

792. "The Privatizing of Eastern Europe," *Institutional Investor* (April 1990): 171.

793. "Managing the Managers," *Reason* (Mar. 1991): 48.

794. "The Privatizing of Eastern Europe," *Institutional Investor* (April 1990): 171; Gerd Schwartz, "Privatization: Possible Lessons from the Hungarian Case," *World Development* 19 (1991): 1731.

795. Ibid.
796. Ibid.
797. "Privatisation in Eastern Europe," *Economist*, April 14, 1990, 19. See Emilia Sebok, "Speeding Up Privatization," *Hungarian Economy* 18, nos. 3-4 (1990): 10, who argue that without speedy privatization, the economy will stagnate.
798. Stark, "Privatization in Hungary: From Plan to Market or from Plan to Clan?" *East European Politics and Societies* 4 (Fall 1990): 351, 388, 392.
799. David M. Newberry, "Reform in Hungary: Sequencing and Privatisation," *European Economic Review* 35 (1991): 571, 576.
800. "Privatization in Hungary," *MTI Econews*, Nov. 21, 1991.
801. "The Privatizing of Eastern Europe," *Institutional Investor* (Apr. 1990): 171.
802. Ibid.
803. Stark, "Privatization in Hungary: From Plan to Market or from Plan to Clan?" *East European Politics and Societies* 4 (Fall 1990): 351, 370-71.
804. Ibid.
805. "Hungary: Need for Western Capital," *Financial Times*, Oct. 30, 1991.
806. Ibid.
807. "On the Back Burner," *Hungarian Observer*, (Oct. 1990).
808. Gerd Schwartz, "Privatization: Possible Lessons from the Hungarian Case," *World Development* 19 (1991): 1731-32.
809. See, for example, "Too Many Firms, Too Few Buyers," *Economist*, Sept. 21, 1991, 14.
810. David M. Newberry, "Reform in Hungary: Sequencing and Privatisation," *European Economic Review* 35 (1991): 571, 576.
811. Ibid.
812. Suzanne Leoffelholz, "Dealing on the Danube," *Financial World*, Mar. 6, 1990, 38, 42.
813. Ibid., 178.
814. Peter A. Bod, "Deregulation and Institution Building—Lessons from the Reform of the Hungarian Public Sector," *Jarbuch der Wirtschaft Osteuropas*, band 13, nr. 1 (1990).
815. Ibid.
816. "Privatisation in Eastern Europe," *Economist*, April 14, 1990.
817. Ibid.
818. By the end of 1992, halfway through the four-year period, only 2.5 to 3 percent of state-owned companies, representing approximately 10 percent of state assets, had been privatized. See "Privatization: Concessions to Domestic Buyers," *MTI Econews*, Dec. 23, 1992; "State to Keep Ownership of 100 Enterprises," *MTI Econews*, Oct. 7, 1992.
819. Emilia Sebok, "Speeding Up Privatization," *Hungarian Economy* 18, nos. 3-4 (1990): 10.
820. There is some reason not to believe these statistics. Much of the Hungarian economy is in private hands, and much of that private economy is unofficial. Government statisticians have not yet developed models to account for such activity and instead rely primarily on the more measurable state output, which everyone acknowledges to be slipping badly. See "Flood of Foreign Investment Capitalizes on the New Hungary," *Washington Post*, Nov. 10, 1991, H1. The gloomy official measurements of the Hungarian economy are also at odds with more circumstantial evidence. For example, in 1991 hard currency deposits in Hungary doubled to approximately $1.5 billion in comparison with figures from the previous year.

821. "Hungary Makes the Most of Economic Head Start," *Chicago Tribune*, Oct. 23, 1991, 1.
822. "Hungary's Output Falls But Privatisation Speeds Up," Reuters, Dec. 5, 1991.
823. *Budapest Week*, Oct. 8–14, 1992, 4. A government spokesperson predicted at the same time that industrial production would increase in 1993 by approximately 3 percent (see p. 2).
824. See supra Chapter Three.
825. "Brake on Hungary's Privatisation Drive," *Times* (London), July 30, 1990, Business Section.
826. "Hungary Unveils First Phase of Major Privatization Program," *BNA Daily Report for Executives*, Sept. 18, 1990.
827. "Hungary's Privatization Program Yields $1.3B, but Results Disappoint," *Global Financial Markets*, Aug. 26, 1991.
828. "Proposal on Privatization Strategies in Hungary," *MTI Econews*, June 3, 1991.
829. In March 1991, the SPA published a document stating that, due to lack of domestic demand, foreign purchases would be encouraged, even though that meant foreign domination of the process. See State Property Agency, "Privatization and Foreign Investment in Hungary (Mar. 1991).
830. "Hungary Puts Speed Above Control in Privatization," *Reuters Money Report*, Jan. 2, 1992.
831. See supra Chapter Five.
832. See supra Chapter Seven.
833. Ibid.
834. Nigel Ash, "The Gearbox Needs an Overhaul," *Euromoney Supplement* (April 1992): 41–42.
835. "Hungary: A Giant Step Ahead," *Business Week*, April 15, 1991, 58.
836. Ibid.
837. "Hungary Unable to Collect Target Revenue Through Privatization," Xinhua General Overseas News Service, Sept. 10, 1991.
838. Ibid.
839. "Hungary's Privatization Program Yields $1.3B, but Results Disappoint," *Global Financial Markets*, Aug. 26, 1991.
840. "Slow Progress in Hungarian Privatization," Agence France Presse, Oct. 15, 1991.
841. "Hungary: Income from Privatisation in First Five Months of 1992," *BBC Summary of World Broadcasts*, July 2, 1992.
842. "Privatization: Concessions to Domestic Buyers," *MTI Econews*, December 23, 1992.
843. "Privatization Revenues Double," *Daily News* (Budapest), Nov. 6–12, 1992, 4.
844. Ibid.
845. "Privatization: Concessions to Domestic Buyers," *MTI Econews*, December 23, 1992.
846. Curiously, Hungarian economist Bela Csikos Nagy argued in 1991 that the market value of Hungarian state assets was actually greater than book value, since book value did not account for intangibles and land values carried on the books at figures far below market value. Bela Csikos Nagy, "Privatization in a Post-Communist World—The Case of Hungary," *Hungarian Business Herald* 2 (1991): 36, 37.

847. "Government Approves Plan to Speed Up Privatization," *MTI Econews*, Dec. 11, 1992. To arrive at that figure, some arithmetic is needed. SPA chief Lajos Csepi is quoted in the article as saying that approximately 100 billion forints worth of state assets had been sold and that this represented approximately 20 percent of all state assets. That would put the total value of state assets to be privatized at 500 billion forints. At 75 forints to the dollar, that works out to $6.7 billion.

848. "Brake on Hungary's Privatization Drive," *Times* (London), July 30, 1990, Business Section.

849. The goal has been variously stated. See, for example, Russell Johnson, "Privatization in Hungary Is Creating Investment Opportunities for U.S. Firms," *Business America*, Jan. 14, 1991, 112, which reports a goal of reducing state ownership of the economy to 30 percent by 1996; "The Privatisation Dilemma," *Euromoney* (Sept. 1990): 145, reports a goal of making the private economy two-thirds of the total economy by 1994; "Hungary Releases Its Privatisation Short-List," *Financial Times*, Sept. 15, 1990, 13, reports a goal of reducing state-owned percentage of economy to 40 percent by 1995.

850. Whereas the government hoped to sell $500 million worth of assets in 1993, it estimated it would lose $867 million worth of assets due to deterioration of their market value. "Government on Economic Issues; SPA on Privatization Speed-Up," *MTI Econews*, Dec. 11, 1992.

851. "The Privatizing of Eastern Europe," *Institutional Investor* (April 1990): 171, 178.

852. See supra notes 194–95, 245–49, and accompanying text.

853. "Delay in Privatization Decisions Reduces Value of Hotel Chains," *MTI Econews*, Nov. 19, 1991.

854. State Property Agency, "First Privatization Program (1990).

855. "Delay in Privatization Decisions Reduces Value of Hotel Chains," *MTI Econews*, Nov. 19, 1991. As of the end of 1992, Pannonia was still owned by the state. HungarHotels was being split into individual properties and sold. The Duna Intercontinental was sold to Marriott for $53.1 million (see "Marriott to Purchase Duna-Intercontinental," *MTI Econews*, Oct. 6, 1992), and the delapidated Grand Hotel Royal was sold to a French consortium for an unbelievable price of $150 million. See "Hotel Sale Epitomizes Hungary's Sometimes Painful Privatization," *Mergers and Acquisitions Report* 5, no. 25 (June 22, 1992): 14. Danubius Hotels was, at the end of 1992, in the middle of a public offering of its stock. A 25 percent stake was sold first, with domestic Hungarians having the first opportunity to purchase on favorable terms. If they did not purchase all the 25 percent, the rest would be sold in Hungary to other buyers. The SPA planned a later sale of a larger stake abroad, perhaps to institutional buyers. See "Hungary Plans Further Public Shares Offering," *Financial Times*, Nov. 25, 1992; "Danubius Share Issue Going Well, *MTI Econews*, Dec. 3, 1992.

856. "Hungary: Privatization Proves to Be a Long, Hard Process," Inter Press Service, May 29, 1992.

857. "Hungarian Entrepreneurs Buy Ailing State-Owned Giant Company," Agence France Presse, Dec. 5, 1991.

858. "Consortium Buys Ailing Videoton," *MTI Econews*, Dec. 6, 1991.

859. "Privatization of Hungarian Banks: State Must Reduce Ownership Share by Law," *MTI Econews*, Nov. 26, 1991.

860. "Government on Economic Issues; SPA on Privatization Speed-Up," *MTI Econews*, Dec. 11, 1992.

861. "Slow Progress in Hungarian Privatisation," Agence France Presse, Oct. 15, 1991.
862. See supra Chapter Three.
863. "Hotel Sale Epitomizes Hungary's Sometimes Painful Privatization," *Mergers and Acquisitions Report* 5, no. 25 (June 22, 1992): 14, reported that the Grand Royal Hotel sold for $150 million; "Marriott to Purchase Duna-Intercontinental," *MTI Econews*, Oct. 6, 1992, states that the Duna-Intercontinental Hotel sold for $53.1 million.
864. "Hungary Acts to Speed Up Flagging Privatisation," *Reuters Library Report*, Dec. 11, 1992; "Government on Economic Issues; SPA on Privatization Speed-Up," *MTI Econews*, Dec. 11, 1992.
865. See supra Chapter Seven.
866. See infra pages 137–138.
867. "Hungary's Privatisation Reorganized," *East European Markets*, Oct. 18, 1991.
868. "Hungary's Privatization Program Yields $1.3B, but Results Disappoint," *Global Financial Markets*, Aug. 26, 1991.
869. "Hungary's Privatisation Reorganized," *East European Markets*, Oct. 18, 1991.
870. "Survey of Privatisation in Eastern Europe," *The Financial Times*, July 3, 1992.
871. Ibid.
872. "Slow Progress in Hungarian Privatisation," Agence France Presse, Oct. 15, 1991.
873. "Hungary Spurs Privatisation As Economic Crisis Looms," *Reuters Library Report*, Nov. 4, 1990.
874. See infra pages 144–145.
875. See infra Chapter Ten.
876. In 1991, 80 percent of sales went to foreigners; in 1992, that figure was 70 percent. See "Government Approves Plan to Speed Up Privatization," *MTI Econews*, Dec. 11, 1992. Other sources have put foreign participation in the years 1990 and 1991 even higher. See, for example, Nigel Ash, "The Gearbox Needs an Overhaul," *Euromoney Supplement* (April 1992): 41, who reports that 90 percent of SPA income came from foreigners; "Hungary Hankers for Native Capital," *Budapest Week*, Sept. 3–9, 1992, 5, states that in 1991, 85 percent of privatization income came from foreigners.
877. "The Privatizing of Eastern Europe," *Institutional Investor* (April 1990); Suzanne Loeffelholz, "Dealing on the Danube," *Financial World*, Mar. 9, 1990, 38–39.
878. "Unhappy Hunting Ground," *Institutional Management* (Sept. 1990).
879. Stark, "Privatization in Hungary: From Plan to Market or from Plan to Clan?" *East European Politics and Societies* 4 (Fall 1990): 351, 361.
880. "The Privatizing of Eastern Europe," *Institutional Investor* (April 1990): 171, 178. This fear of foreign economic hegemony was reportedly not as bad in Hungary as in Poland and Czechoslovakia, which had experienced some unpleasantness with Germans earlier in the century. "Hungary Hankers for Native Capital," *Budapest Week*, Sept. 3–9, 1992, 5.
881. "Hungary Hankers for Native Capital," *Budapest Week*, Sept. 3–9, 1992, 5.
882. "Hungary Readies Privatisation 'Credit Cards'" *Reuters Library Report*, Oct. 12, 1992.

Notes

883. "Government Gives Boost to Home Grown Capitalism," *Financial Times*, May 29, 1992.
884. "Privatisation in Eastern Europe," *Economist*, April 14, 1990; "Unhappy Hunting Ground," *International Management* (Sept. 1990).
885. "The Privatisation Dilemma," *Euromoney* (Sept. 1990): 145, 148.
886. "The Privatizing of Eastern Europe," *Institutional Investor* (Apr. 1990): 171, 175-78.
887. See supra pages 141-144.
888. "Unhappy Hunting Ground," *International Management* (Sept. 1990): 61, 63; "East Europe's Sale of the Century," *New York Times*, May 22, 1990, D1, col. 3.
889. "Proposal on Privatization Strategies in Hungary," *MTI Econews*, June 3, 1991.
890. "East Europe's Sale of the Century," *New York Times*, May 22, 1990, D1, col. 3, puts the figure at more than 20 percent; "Putting Hungary on the Market," *Independent*, Feb. 12, 1990, 23, puts the figure at 30 percent.
891. "Unhappy Hunting Ground," *International Management* (Sept. 1990): 61, 63.
892. "Putting Hungary on the Market," *Independent*, Feb. 12, 1990, 23. See also Bela Csikos Nagy, "Privatization in a Post-Communist Society—The Case of Hungary," *Hungarian Business Herald* 2 (1991): 36, 38; Nagy states that 50 percent of Austrian trade and 40 percent of Austrian industry is controlled by foreigners.
893. Ibid.
894. "Proposal on Privatization Strategies in Hungary," *MTI Econews*, June 3, 1991.
895. State Property Agency, "Privatization and Foreign Investment in Hungary (Mar. 1991): 5.
896. Stark, "Privatization in Hungary: From Plan to Market or from Plan to Clan?," *East European Politics and Societies* 4 (Fall 1990): 351, 358-59.
897. "Hungary: Need for Western Capital," *Financial Times*, Oct. 30, 1991.
898. "Not Selling Out but Looking for Partners," *EGISZ* (July-Sept. 1989).
899. See supra Chapter Seven.
900. See "Government Gives Boost to Home Grown Capitalism," *Financial Times*, May 29, 1992.
901. See supra Chapter Five.
902. "Hungarian Entrepreneur to Buy Pharmatrade," *MTI Econews*, Nov. 13, 1991.
903. Ibid.
904. "Mine Privatized in Hungary," Xinhua General Overseas News Service, Nov. 2, 1991.
905. "First Private Coal Mine in Hungary," *MTI Econews*, Oct. 31, 1991; "Budapest Sells Off Mine," *Wall Street Journal*, Nov. 4, 1991, A10, col. 3.
906. "Government Approves Plan to Speed Up Privatization," *MTI Econews*, Dec. 11, 1992.
907. "Steady Growth in Personal Savings," *MTI Econews*, Sept. 17, 1992.
908. See supra Chapter Three.
909. See Chapter Three. The compensation legislation is discussed extensively supra Chapter Two.
910. "Hungary Acts to Speed Up Flagging Privatisation," *Reuters Library Report*, Dec. 11, 1992.
911. See supra Chapter Four.

912. Act VI of 1990 on Securities and the Stock Exchange, which provided the legal basis for opening the exchange, was passed by the Parliament in January 1990. The SPA began its work in March of that year, and the Budapest Stock Exchange opened formally in June 1990.
913. "Hungary: Need for Western Capital," *Financial Times*, Oct. 30, 1991.
914. "Industry Minister Interviewed on Privatisation," *BBC*, June 23, 1990.
915. "First Privatization Programme Dragging On," *MTI Econews*, Aug. 27, 1991.
916. "Flood of Foreign Investment Capitalizes on the New Hungary," *Washington Post*, Nov. 10, 1991, H1.
917. "Privatization in Hungary—an Econews Background," *MTI Econews*, April 25, 1991.
918. See supra Chapter Three.
919. "First Privatization Programme Dragging On," *MTI Econews*, Aug. 27, 1991.
920. "Calling for the Next Stage in Property Rights; An Interview with Dr. Lajos Csepi," *Privat Profit* (Sept. 1991).
921. "The Painful Road to Privatization," *Financial Times*, Nov. 5, 1991, I29.
922. "On the Stock Exchange's Thorny Path," *Hungarian Observer* (Mar. 1992).
923. "Plans for Compensation Certificate-for-Share Exchange," *MTI Econews*, Dec. 9, 1991. According to one estimate, the over-the-counter market for Hungarian securities represents approximately 90 percent of the total trading volume in the country. Interview with Dr. Mariann Vida, 1991.
924. "Flood of Foreign Investment Capitalizes on the New Hungary," *Washington Post*, Nov. 10, 1991, H1; "Hungary: Need for Western Capital," *Financial Times*, Oct. 30, 1991.
925. "The Painful Road to Privatization," *Financial Times*, Nov. 5, 1991, I29.
926. See supra Chapter Three.
927. "Sale of Hotel Chain Postponed After Proposed Change in Tax Break," *BNA International Business Daily*, Nov. 20, 1991.
928. Ibid.
929. Nigel Ash, "The Gearbox Needs an Overhaul," *Euromoney Supplement* (Apr. 1992): 41.
930. "On the Stock Exchange's Thorny Path," *Hungarian Observer* (Mar. 1992).
931. "Plans for Compensation Certificate-for-Share Exchange," *MTI Econews*, Dec. 9, 1991.
932. See supra Chapter Seven.
933. See Chapter Seven.
934. "Putting Hungary on the Market," *Independent*, Feb. 12, 1990, 23.
935. "Eastern Europe—Early Days," *Euromoney Supplement* (July 1990): 2.
936. See supra Chapter Three.
937. "Eastern Europe—Early Days," *Euromoney Supplement* (July 1990): 2; "Flood of Foreign Investment Capitalizes on New Hungary," *Washington Post*, Nov. 10, 1991, H1.
938. "Hungary Spurs Privatisation As Economic Crisis Looms," *Reuter Library Report*, Nov. 4, 1990.
939. Ibid.
940. "Proposal on Privatization Strategies in Hungary," *MTI Econews*, June 3, 1991.

Notes

941. "Economic Cabinet Discusses Ministry of Finance Privatization Strategy," *MTI Econews*, June 18, 1991.
942. Ibid.
943. The Self-Privatization Program is discussed extensively above. See supra Chapter Five.
944. See infra Chapter Ten.
945. Nigel Ash, "The Gearbox Needs an Overhaul," *Euromoney Supplement* (Apr. 1992): 41.
946. Ibid.
947. Gerd Schwartz, "Privatization: Possible Lessons from the Hungarian Case," *World Development* 19 (1991): 1731-34.
948. This problem is discussed supra Chapter Three and infra Chapter Ten.
949. "Pick Salami Shares Aimed at Small Investors," *MTI Econews*, July 16, 1992.
950. "Public Issue of Pick Salami Shares to Start October 27," *MTI Econews*, Sept. 30, 1992.
951. "Danubius Privatization—Further Details," *MTI Econews*, Sept. 22, 1992.
952. State Property Agency, "Privatization and Foreign Investment in Hungary" (Mar. 1991): 8.
953. "Proposal on Privatization Strategies in Hungary," *MTI Econews*, June 3, 1991.
954. "Kupa on Unemployment, Central Budget," *MTI Econews*, Aug. 6, 1991.
955. "Government to Offer 600 More State Owned Properties for Sale," *BNA International Business Daily*, Feb. 6, 1992.
956. "Income From Privatization in Hungary," *MTI Econews*, Sept. 9, 1991.
957. "1992 Budget Income From Privatization," *MTI Econews*, Sept. 23, 1991.
958. "Government to Offer 600 More State Owned Properties for Sale," *BNA International Business Daily*, Feb. 6, 1992.
959. "Privatization Revenues Top HuF 26B," *MTI Econews*, May 27, 1992.
960. A scandal erupted in early 1993 when it was discovered that the Antall government was hiding government money in secret foundations headed by prominent MDF members. The implication was that the MDF was appropriating public funds to support its reelection efforts in 1994. See *Budapest Sun*, April 22, 1993, 1.
961. "Hungary Lists Firms That Will Not Be Privatized," *Reuters Money Report*, May 28, 1991.
962. Ibid.
963. "Hungarians Need Bigger Role in Privatisation—SPA," *Reuters Money Report*, Jan. 13, 1992.
964. "Economic Affairs in Brief," *BBC Summary of World Broadcasts*, Jan. 31, 1992; "Government Expects 79% of State Assets Will Be Privatised," *BBC Summary of World Broadcasts*, Feb. 12, 1992.
965. "Hungary State Holding Co. May Speed Privatisation," *Reuters Money Report*, Jan. 21, 1992.
966. "Hungary State Holding Co. May Speed Privatisation," *Reuters Money Report* (Jan. 21, 1992) quotes Tamas Szabo, minister without portfolio for Privatization.
967. "Hungarian Parliament Approves New Agency to Manage Enterprises Retained by State," *BNA International Business Daily*, June 24, 1992.
968. "Companies to Retain State Interest," *MTI Econews*, June 24, 1992.

969. Ibid.
970. "State to Keep Ownership of 100 Enterprises," *MTI Econews*, Oct. 7, 1992.
971. "Planned Disposal of Assets of State Property Agency," *BBC Summary of World Broadcasts*, Aug. 6, 1992.
972. "State to Keep Ownership of 100 Enterprises," *MTI Econews*, Oct. 7, 1992.
973. "Hungary's Szabo Gives His Reasons for Leaving SPA," *East European Markets*, Jan. 10, 1992.
974. "Privatization Bills," *MTI Econews*, Mar. 27, 1992.
975. "Hungary's Big Sell-Off Hits Hard Times," *Financial Times*, May 21, 1992, I2, quotes Karoly Szabo, deputy managing director of the SPA.
976. See supra Chapter Nine.
977. "Hungary Readies Privatisation 'Credit Cards'" *Reuter Library Report*, Oct. 12, 1992.
978. David Fairlamb, "The Privatizing of Eastern Europe," *Institutitonal Investor* (April 1990): 171; "East Europe's Sale of the Century," *New York Times*, May 22, 1990, D1, col. 3; Suzanne Loeffelholz, "Dealing on the Danube," *Financial World*, March 9, 1990, 38.
979. Stark, "Privatization in Hungary: From Plan to Market or from Plan to Clan?" *East European Politics and Societies*" 4 (Fall 1990): 351, 370–71.
980. "Hungary: Need for Western Capital," *Financial Times*, Oct. 30, 1991.
981. "Hungary Hankers for Native Capital," *Budapest Week*, Sept. 3–9, 1992, 5, which quotes George Hollo of SPA.
982. "Steady Growth in Personal Savings," *MTI Econews*, Sept. 17, 1992.
983. "Government Approves Plan to Speed Up Privatization," *MTI Econews*, Dec. 11, 1992.
984. "Privatization: Concessions to Domestic Buyers," *MTI Econews*, December 23, 1992.
985. "The Privatizing of Eastern Europe," *Institutional Investor* (April 1990): 171, 173.
986. "On the Back Burner," *Hungarian Observer* (Oct. 1990).
987. David Fairlamb, "The Privatizing of Eastern Europe," *Institutional Investor* (April 1990): 171–72.
988. "Foreign Investment in Privatization," *MTI Econews*, Aug. 11, 1992.
989. Hungary began in 1990 by focusing on the "plums" rather than the "lemons" in a conscious attempt to put the privatization program on a sound footing. See "East Europe's Sale of the Century," *New York Times*, May 22, 1990, D1, col. 3.
990. At the same time the government was instituting measures to spur domestic demand, total privatization revenues were expected to drop approximately 36 percent from 1992 to 1993. "Privatization: Concessions to Domestic Buyers," *MTI Econews*, Dec. 23, 1992.
991. David Fairlamb, "The Privatizing of Eastern Europe," *Institutional Investor* (April 1990): 171.
992. Suzanne Loeffelholz, "Dealing on the Danube," *Financial World*, March 6, 1990.
993. "Managing the Managers," *Reason* (Mar. 1991): 48.
994. See supra chapters Four and Five.
995. See supra Chapter Four.
996. See supra Chapter Seven.
997. See, for example, David Fairlamb, "The Privatizing of Eastern Europe," *Institutional Invester* (April 1990): 171; Stark, "Privatization in Hungary: From Plan

to Market or from Plan to Clan?" *East European Politics and Societies* 4 (Fall 1990): 351; "The Privatisation Dilemma," *Euromoney* (Sept. 1990): 145; Suzanne Loeffelholz, "Dealing on the Danube," *Financial World*, March 6, 1990; "Hungary's Big Sell-Off Hits Hard Times," *Financial Times*, May 21, 1992, I2; Joseph Bell, "Privatization in Central and Eastern Europe," *PLI Course Handbook*, Sept. 25-26, 1991; "Challenges of Privatization Keep the Market Busy," *Chemical Week*, March 18, 1992, 24.

998. See supra chapters Three and Four.
999. "On the Back Burner," *The Hungarian Observer* (Oct. 1990).
1000. "Unhappy Hunting Ground," *International Management* (Sept. 1990): 61.
1001. See, for example, Nigel Ash, "The Gearbox Needs an Overhaul," *Euromoney Supplement* (April 1992): 41, 49; "Head of U.S. Investment Fund Criticizes Slow Pace of Hungarian Privatization," *BNA International Trade Reporter*, Feb. 19, 1992, 322, which quotes Alexander C. Tomlinson, president of Hungarian American Enterprise Fund; "Survey of Privatisation in Eastern Europe," *Financial Times*, July 3, 1992.
1002. "Survey of Privatisation in Eastern Europe," *Financial Times*, July 3, 1992.
1003. "Hungary's Little Shop of Dreams," *Washington Post*, May 28, 1991, E1, col. 3.
1004. Suzanne Loeffelholz, "Dealing on the Danube," *Financial World*, March 6, 1990, 38-39.
1005. "Survey of Privatisation in Eastern Europe," *Financial Times*, July 3, 1992.
1006. See supra Chapter Three.
1007. See supra chapters Three and Four.
1008. See "Hungary Spurs Privatisation As Economic Crisis Looms," *Reuters Library Report*, Nov. 4, 1990.
1009. See supra Chapter Six.
1010. "Too Many Firms, Too Few Buyers," *Economist*, Sept. 21, 1991, 14.
1011. "Government to Offer 600 More State-Owned Properties For Sale," *BNA International Business Daily*, Feb. 6, 1992.
1012. Interview with Drs. Zsolt Sztrokay and Zsuzsanna Zatik, April 27, 1993.
1013. "Rightist Manifesto Causes Furor," *Facts on File World News Digest*, Nov. 12, 1992.
1014. See supra Chapter Three.
1015. See supra Chapter Six.
1016. See, for example, "Hungary's Little Shops of Dreams," *Washington Post*, May 28, 1991, E1, col. 3; "Hungary's Big Sell-Off Hits Hard Times," *Financial Times*, May 21, 1992, I2; Nigel Ash, "The Gearbox Needs an Overhaul," *Euromoney Supplement* (April 1992): 41; "Government to Offer 600 More State-Owned Properties for Sale," *BNA International Business Daily*, Feb. 6, 1992; "Few Privatizations in the East Bloc Go Very Quickly," *National Law Journal*, Feb. 24, 1992, 19.
1017. "Hungary: A Race Against Time," *Financial Times*, Oct. 30, 1991.
1018. "Last Breath of the Pickled Cucumber," *Times* (London), Nov. 21, 1991.
1019. "Public Placement of Danubius Hotel Shares," *MTI Econews*, July 16, 1992.
1020. The protracted process of converting a state enterprise into a share company has often been observed to be a problem slowing down privatization. See, for example, "New Privatization Program Launched to Break Up 'Hollow' Parent

Companies," *BNA International Business Daily*, April 16, 1992, which quotes an American lawyer working in Budapest. The unclear ownership structures explain much of the delay.

1021. See, for example, David Fairlamb, "The Privatizing of Eastern Europe," *Institutional Investor* (April 1990): 171; "Privatization in Eastern Europe," *Economist*, April 14, 1990, 19-20.

1022. Peter A. Bod, "Deregulation and Institution Building—Lessons from the Reform of the Hungarian Public Sector," *Jarbuch der Wirtschaft Osteuropas*, band 13, nr. 1 (1990).

1023. David Bartlett, "The Political Economy of Privatization: Property Reform and Democracy in Hungary," *East European Politics and Societies* (Winter 1992): 73, 91; "Time to Sort Out Who Owns What," *Financial Times*, April 16, 1991, I18; Speech of Sir William Ryrie in Czechoslovakia, July 1991.

1024. "Hungary Hankers for Native Capital," *Budapest Week*, Sept. 3-9, 1992, 5.

1025. Suzanne Loeffelholz, "Dealing on the Danube," *Financial World*, March 6, 1992, 38, 42. The government did move in 1991 to rectify this lack. See, for example, Act XXXVIII of 1991 on Registered Design Patents, III HRLF 2 (1992); Act XXXIX of 1991 on the Patent of the Topography of Microelectronic Semi-Conductor Products, III HRLF 2 (1992).

1026. "Strong Domestic Interest and Controversy Provoked by Hungary's Privatisation Policy," *Financial Times*, Sept. 20, 1990, I30; Suzanne Loeffelholz, "Dealing on the Danube," *Financial World,"* March 6, 1990, 38; "New Privatization Program Launched to Break Up 'Hollow' Parents," *BNA International Business Daily*, April 16, 1992.

1027. "The Privitisation Dilemma," *Euromoney* (Sept. 1990): 145-46.

1028. Ibid.

1029. David Bartlett, "The Political Economy of Privatization: Property Reform and Democracy in Hungary," *East European Politics and Societies* 6 (Winter 1992): 73, 91; "Strong Domestic Interest and Controversy Provoked by Hungary's Privatisation Policy," *Financial Times*, Sept. 20, 1990, I30; Suzanne Loeffelholz, "Dealing on the Danube," *Financial World"*, March 6, 1990, 38.

1030. "Time to Sort Out Who Owns What," *Financial Times*, April 16, 1991, I18.

1031. See supra Chapter Two.

1032. David Bartlett, "The Political Economy of Privatization: Property Reform and Democracy in Hungary," *East European Politics and Societies* (Winter 1992): 73, 91.

1033. See supra Chapter Nine.

1034. "Hungary Spurs Privatisation As Economic Crisis Looms," *Reuters Library Report*, Nov. 4, 1990.

1035. "Hungary's Privatization Program Yields $1.3B, but Results Disappoint," *Global Financial Markets*, Aug. 26, 1991.

1036. "Government to Offer 600 More State-Owned Properties for Sale," *BNA International Business Daily*, Feb. 6, 1992.

1037. See supra Chapter Three.

1038. See supra page 158.

1039. See supra Chapter Seven.

1040. "SPA Strikes BZW Off Its Adviser List," *Daily News* (Budapest), Nov. 6-12, 1992, 4.

1041. Ibid.

1042. "SPA Bounces British Bank in Beer Brawl," *Budapest Week*, Nov. 12-18, 1992, 2.

1043. Ibid.

1044. "Barclays and SPA Clarify Differences," *Daily News* (Budapest), Nov. 13-19, 1992, 4.

1045. David Fairlamb, "The Privatizing of Eastern Europe," *Institutional Investor* (April 1990): 171, 173.

1046. "East Europe's Sale of the Century," *New York Times*, May 22, 1990, D1, col. 3.

1047. David Fairlamb, "The Privatizing of Eastern Europe," *Institutional Investor* (April 1990): 171, 173.

1048. "Hungary's Big Sell-Off Hits Hard Times," *Financial Times*, May 21, 1992, I2.

1049. "Vagaries of Valuing Assets Pose Problems for Investors," *BNA International Business Daily*, July 27, 1992, describes overstaffing in the hotel industry.

1050. The deterioration was particularly acute among those firms that depended on the Soviet Union as a principal market. See "Hungary's Big Sell-Off Hits Hard Times," *Financial Times*, May 21, 1992, I2. Overall, at the end of 1992, state firms were losing value at the rate of approximately $72.3 million per month. See "Government on Economic Issues; SPA on Privatization Speed-Up," *MTI Econews*, Dec. 11, 1992.

1051. "Hungary: Privatization Proves to Be a Long, Hard Process," Inter Press Service, May 29, 1992.

1052. "Survey of Privatisation in Eastern Europe," *Financial Times*, July 3, 1992.

1053. See supra Chapter Three.

1054. Law LXII of 1992 on Telecommunications.

1055. Ibid., sec. 4.

1056. Debevoise and Plimpton, "Central and Eastern European Bulletin," March 23, 1993.

1057. Law XXV of 1991 on Partial Compensation for Damages Unlawfully Caused by the State to Properties Owned by Citizens in the Interest of Settling Ownership Relations.

1058. See "Property Compensation Law to Take Effect in Hungary," *BNA International Trade Reporter* 8, no. 32 (1991): 1185.

1059. The preamble to the act states as one goal "settling ownership relations and bringing about safety for enterprises necessary for creating a market economy." Law XXV of 1991 on Partial Compensation for Damages Unlawfully Caused by the State to Properties Owned by Citizens in the Interest of Settling Ownership Relations.

1060. See "Legislation on Privatization and Acquisitions," *Doing Business with Eastern Europe*, Aug. 1, 1991.

1061. Law XXV of 1991 on Partial Compensation for Damages Unlawfully Caused by the State to Properties Owned by Citizens in the Interest of Settling Ownership Relations, secs. 13-28.

1062. On this subject, see generally "Hungary: Privatization and the Environment, *Business Eastern Europe* 20, April 20, 1992, 189; "The Greening of Central and Eastern Europe," *Legal Times*, Dec. 7, 1992, 35.

1063. Several advisers counseling would-be bidders on privatized companies cautioned that the costs of environmental cleanup should be factored into the decision-making process. See, for example, Joseph Bell, "Privatization in Central and Eastern Europe," *PLI Course Handbook*, Sept. 25–26, 1991; speech of Sir William Ryrie, Czechoslovakia, July 1991.

Bibliography

Acts of the Hungarian Parliament

Company Law of 1875
Act VI of 1959 on the Civil Code of the Peoples' Republic of Hungary
Act VI of 1988 on Business Organizations
Act XXIV of 1988 on Investments of Foreigners in Hungary
Act XIII of 1989 on the Conversion of Economic Organizations and Business Associations
Act LXXIV of 1990 on Privatization of State Enterprises Dealing with Detail Trade, Catering Trade, and Consumer Services
Act LXXXVI of 1990 on the Prohibition of Unfair Market Practices
Act VII of 1990 on Foundation of State Property Agency with the Purpose of the Management and Utilization of Property Pertaining to This
Act LXXIV of 1990 on the Privatization (Alienation, Utilization) of State-Owned Companies Engaged in Retail Trade, Catering, and Consumer Services
Act VI of 1990 on Securities and the Stock Exchange
Act VIII of 1990 on the Protection of Property Entrusted to Enterprises of the State
Act LXIX of 1991 on Banks and Banking Activities
Act XXXVIII of 1991 on Registered Design Patents
Act IL of 1991 on Bankruptcy Procedures, Liquidation Procedures, and Final Settlement
Act LXIII of 1991 on Investment Funds
Act XXXIX of 1991 on the Patent of the Topography of Microelectronic Semi-Conductor Products
Act LX of 1991 on the National Bank of Hungary
Act XVIII of 1991 on Accounting
Act XXV of 1991 on Partial Compensation for Damages Unlawfully Caused by the State to Properties Owned by Citizens in the Interest of Settling Ownership Relations.
Act LIII of 1992 on the Management and Utilization of Entrepreneurial Assets Permanently Remaining in State Ownership
Act LXII of 1992 on Telecommunications
Act LIV of 1992 on the Sale, Utilization, and Protection of Assets Temporarily Owned by the State
Act XLIV of 1992 on the Employees' Part-Ownership Program

News Stories

"Not Selling Out but Looking for Partners." *EGISZ* (July–Sept. 1989).
Sebok, Emilia. "Speeding Up Privatization." *Hungarian Economy* 18, nos. 3–4 (1990): 10.
"Putting Hungary on the Market." *Independent*, Feb. 12, 1990, 23.
Loeffelholz, Suzanne. "Dealing on the Danube." *Financial World*, March 6, 1990, 42.
"Hungary Seeks UK Help on Privatisation." *Financial Times*, March 8, 1990, sec. 1, p. 3.
"Privatization in Eastern Europe." *Economist*, April 14, 1990.
Fairlamb, David. "The Privatizing of Eastern Europe." *Institutional Investor* (April 1990): 171.
"East Europe's Sale of the Century." *New York Times*, May 22, 1990, D1, col. 3.
"Hungary Hurries to Privatisation." *Times* (London), June 5, 1990, Business Section.
"Industry Minister Interviewed on Privatisation." BBC, June 23, 1990.
"Hungarian State Body Attacked on Privatisation." *Financial Times*, July 6, 1990, I2.
"Privatisation Row Erupts in Hungary." *Independent*, July 6, 1990, 21.
"Hungarian Companies on the Block Are Named." *Financial Times*, July 13, 1990, I25.
"Dispute in Hungary Privatization." *Business Eastern Europe*, July 16, 1990, 235.
"Hungary Replaces Criticized Privatization Chief." *Reuters Library Report*, July 26, 1990.
"Brake on Hungary's Privatisation Drive." *Times* (London), July 30, 1990, Business Section.
"Eastern Europe—Early Days." *Euromoney Supplement: The Global Sweep of Privatization* (July 1990): 2.
"In Eastern Europe, the Big Sell-Off Is Set to Begin." *Business Week*, Aug. 6, 1990, 42.
"Privatization in Hungary: Official Lists Not Needed." *Business Eastern Europe*, Aug. 13, 1990, 267.
"Hungary." *International Reports*, Sept. 7, 1990.
"Hungary Releases Its Privatisation Short-List." *Financial Times*, Sept. 15, 1990, I3.
"Hungary Unveils First Phase of Major Privatization Program." *BNA Daily Report for Executives*, Sept. 18, 1990.
"Strong Domestic Interest and Controversy Provoked by Hungary's Privatisation Policy." *Financial Times*, Sept. 20, 1990, I30.
"The Privatisation Dilemma." *Euromoney* (Sept. 1990):145–46.
"Unhappy Hunting Ground." *International Management* (Sept. 1990): 63.
"Hungary's Industry Up for Sale." *Business Law Brief, Financial Times* (Oct. 1990).
"On the Back Burner." *Hungarian Observer* (Oct. 1990).
"Hungary Spurs Privatisation As Economic Crisis Looms." *Reuters Library Report*, Nov. 4, 1990.
"Free Market Changes Underway." *Hungary Handbook* (Nov. 1990).
"Acquisitions in the East: Europe's New M & A Frontier." *M & A Europe* (Nov.–Dec. 1990).
"Property Compensation Law to Take Effect in Hungary." *BNA International Trade Reporter* 8, no. 32 (1991): 1185.

Johnson, Russell. "Privatization in Hungary Is Creating Investment Opportunities for U.S. Firms." *Business America*, January 14, 1991, 112.
"Off on the Wrong Foot." *Hungarian Observer* (Jan. 1991).
"Privatization Loan in Hungary." *MTI Econews*, Feb. 8, 1991.
"Hotel Chain to Come Under the Privatization Hammer." *MTI Econews*, Feb. 13, 1991.
"Managing the Managers." *Reason* (Mar. 1991): 48.
"Hungary: A Giant Step Ahead." *Business Week*, April 15, 1991, 58.
"Time to Sort Out Who Owns What." *Financial Times*, April 16, 1991, I18.
"Hungary's Budding Entrepreneurs Bid for Shops and Cafes." *Reuters Library Report*, April 22, 1991.
"Privatization in Hungary—an ECONEWS Background." *MTI Econews*, April 25, 1991.
"Privatization in Hungary—Konrad Adenauer Foundation." *MTI Econews*, May 27, 1991.
"Hungary's Little Shops of Dreams." *Washington Post*, May 28, 1991, E1, col. 3.
"Hungary Lists Firms That Will Not Be Privatized." *Reuters Money Report*, May 28, 1991.
"Proposal on Privatization Strategies in Hungary." *MTI Econews*, June 3, 1991.
"State Property Agency to Announce New Asset Handling Tender." *MTI Econews*, June 12, 1991.
"Economic Cabinet Discusses Minister of Finance Privatization Strategy." *MTI Econews*, June 18, 1991.
"National Bank of Hungary's Role in Hotel Company Privatization." *MTI Econews*, June 24, 1991.
"SPA Invites Consultancy Bids for Self-Privatisation Programme." *MTI Econews*, July 8, 1991.
"Enthusiasm for Foreign Investment in Hungary Fails to Tell Story." *East European Markets*, July 12, 1991. "Latest SPA Privatization Tender Attracts 200 Consultancy Firms." *MTI Econews*, July 23, 1991.
"Kupa on Unemployment, Central Budget." *MTI Econews*, Aug. 6, 1991.
"Self-Privatisation to Start in Hungary Soon." *MTI Econews*, Aug. 6, 1991.
"Hungary's Privatization Program Yields $1.3B, but Results Disappoint." *Global Financial Markets*, Aug. 26, 1991.
"First Privatization Programme Dragging On." *MTI Econews*, Aug. 27, 1991.
"Sachs, The Property Distributor." *Privat Profit* (Aug. 1991).
"An Open Letter About Property." *Privat Profit* (Aug. 1991).
"Income from Privatization in Hungary." *MTI Econews*, Sept. 9, 1991.
"Hungary Unable to Collect Target Revenue Through Privatization." Xinhua General Overseas News Service, Sept. 10, 1991.
"Too Many Firms, Too Few Buyers." *Economist*, Sept. 21, 1991, 14.
"1992 Budget Income from Privatization." *MTI Econews*, Sept. 23, 1991.
"Private Possibilities." *Privat Profit* (Sept. 1991).
"Calling for the Next Stage in Property Rights; An Interview with Dr. Lajos Csepi." *Privat Profit* (Sept. 1991).
"World Economy 6; Focus on Privatisation in Eastern Europe." *Financial Times*, Oct. 14, 1991.
"Slow Progress in Hungarian Privatisation." Agence France Presse, Oct. 15, 1991.
"Hungary's Privatisation Reorganized." *East European Markets*, Oct. 18, 1991.
"Eastern Europe Tries to Stoke Up Its Fire Sale." *Business Week*, Oct. 21, 1991, 52.

"Hungary Makes the Most of Economic Head Start." *Chicago Tribune*, Oct. 23, 1991, 1.
"Danubius Privatization Called Off." *MTI Econews*, Oct. 28, 1991.
"Hungarian Hotels Sale Delayed by Six Months." *Financial Times*, Oct. 29, 1991.
"Hungary's Privatization Drive in Disarray." Reuters, Oct. 29, 1991.
"Fumbled Hotel Sale Snags Hungary's Privatization Scheme." *Reuters Library Report*, Oct. 29, 1991.
"Hungary: Need for Western Capital." *Financial Times*, Oct. 30, 1991
"Hungary: A Race Against Time." *Financial Times*, Oct. 30, 1991.
"First Private Coal Mine in Hungary." *MTI Econews*, Oct. 31, 1991.
"Banking on Privatization: Commercial Banks Are Next in Line to Go Private." *Hungarian Observer*, Oct. 1991.
"Mine Privatized in Hungary." *Xinhua General Overseas News Service*, Nov. 2, 1991.
"Budapest Sells Off Mine." *Wall Street Journal*, Nov. 4, 1991, A10, col. 3.
"The Painful Road to Privatisation." *Financial Times*, Nov. 5, 1991, I29.
"Flood of Foreign Investment Capitalizes on New Hungary." *Washington Post*, Nov. 10, 1991, H1.
"Hungarian Entrepreneur to Buy Pharmatrade." *MTI Econews*, Nov. 13, 1991.
"Delay in Privatization Decisions Reduces Value of Hotel Chains." *MTI Econews*, Nov. 19, 1991.
"Sale of Hotel Chain Postponed After Proposed Change in Tax Break." *BNA International Business Daily*, Nov. 20, 1991.
"Last Breath of the Pickled Cucumber." *Times* (London), Nov. 21, 1991.
"State Property Agency to Sell Its IBUSZ Shares." *MTI Econews*, Nov. 21, 1991.
"Privatization in Hungary." *MTI Econews*, Nov. 21, 1991.
"Privatization of Hungarian Banks: State Must Reduce Ownership Share by Law." *MTI Econews*, Nov. 26, 1991.
"Hungarian Entrepreneurs Buy Ailing State-Owned Giant Company." Agence France Presse, Dec. 5, 1991.
"Hungary's Output Falls but Privatisation Speeds Up." Reuters, Dec. 5, 1991.
"Consortium Buys Ailing Videoton." *MTI Econews*, Dec. 6, 1991.
"Plans for Compensation Certificate-for-Share Exchange." *MTI Econews*, Dec. 9, 1991.
"Hungary Privatisation Programme Hits Crucial Phase." *Reuters Money Report*, Dec. 20, 1991.
"Hungarian Firms Seek Partners." *Eastern European Report*, Dec. 23, 1991.
"Hungary Puts Speed Above Control in Privatization." *Reuters Money Report*, Jan. 2, 1992.
"Hungary's Szabo Gives His Reasons For Leaving SPA." *East European Markets*, Jan. 10, 1992.
"State Liquidation Agency Becomes PLC." *MTI Econews*, Jan. 13, 1992.
"Hungarians Need Bigger Role in Privatisation—SPA." *Reuters Money Report*, Jan. 13, 1992.
"Privatization of Hungarian Commercial Banks." *MTI Econews*, Jan. 14, 1992.
BBC Summary of World Broadcasts, Jan. 18, 1992.
"Hungary State Holding Co. May Speed Privatisation." *Reuters Money Report*, Jan. 21, 1992.
"SPA 'Sells' Hungary's Self-Privatization Programme in London." *MTI Econews*, Jan. 24, 1992.
"Spokesman's Briefing—Loans for Privatisation," *MTI Econews*, Jan. 24, 1992.

"Existence Credit Becomes Cheaper." *MTI Econews*, Jan. 24, 1992.
"Economic Affairs in Brief." *BBC Summary of World Broadcasts*, Jan. 31, 1992.
"Hungarian Banks Could Go On Sale This Year." *Reuters Money Report*, Feb. 5, 1992.
"Government to Offer 600 More State-Owned Properties for Sale." *BNA International Business Daily*, Feb. 6, 1992.
"Government Expects 79% of State Assets Will Be Privatised." *BBC Summary of World Broadcasts*, Feb. 12, 1992.
"World Bank Loan to Hungary." *MTI Econews*, Feb. 15, 1992.
"Head of U.S. Investment Fund Criticises Slow Pace of Hungarian Privatization." *BNA International Trade Reporter*, Feb. 19, 1992, 322.
"Hungarian Banks Expected to Be for Sale During Second Half of This Year." *Eastern Europe Report*, Feb. 24, 1992.
"Few Privatizations in the East Bloc Go Very Quickly." *National Law Journal*, Feb. 24, 1992, 19.
"HAFE to Be Privatized." *MTI Econews*, Feb. 25, 1992.
"Danubius Hotel Chain's Value Revised Downward." *MTI Econews*, Feb. 26, 1992.
"Way Cleared for IBUSZ Privatization." *BBC Summary of World Broadcasts*, Feb. 29, 1992.
"State Assets Agency Launches Second Phase of Privatization." *BBC Summary of World Broadcasts*, Mar. 14, 1992.
"Challenges of Privatization Keep the Market Busy." *Chemical Week*, Mar. 18, 1992, 24.
"Hungary Plows New Territory: State Farms Follow a Bumpy Road to Privatization." *Chicago Tribune*, Mar. 20, 1992.
"Privatization Bills." *MTI Econews*, Mar. 27, 1992.
"On the Stock Exchange's Thorny Path." *The Hungarian Observer* (Mar. 1992).
"Privatization Loan Credits Modified." *MTI Econews*, April 9, 1992.
"New Privatization Program Launched to Break Up 'Hollow' Parents." *BNA International Business Daily*, April 16, 1992.
"Bank Privatization Committee Established." *MTI Econews*, April 16, 1992.
"Hungary Pressed to Privatise." *East European Markets*, April 17, 1992.
"Hungary: Privatization and the Environment." *Business Eastern Europe* 20 (April 20, 1992): 189.
Ash, Nigel. "The Gearbox Needs an Overhaul." *Euromoney Supplement* (April 1992): 41–42.
"Hungary Expands Streamlined Privatisation Scheme." *Reuters Money Report*, May 5, 1992.
"Hungary Eases Sale of State Companies." *Financial Times*, May 6, 1992.
"Simplified Privatization—A City of London View." *MTI Econews*, May 7, 1992.
"Hungary: Uses of Compensation Coupons." *BBC Summary of World Broadcasts*, May 7, 1992.
"New Directorate and Supervisory Committee Appointed for the Credit Bank." *BBC Summary of World Broadcasts*, May 8, 1992.
"Co-Nexus Active in Managing SPA Portfolio." *MTI Econews*, May 13, 1992.
"Czechoslovakia, Hungary Choose Differing Routes to Reach Same Goal—Privatization." *EBRD Watch* 2, no. 19 (May 18, 1992): 4.
"Hungary's Big Sell-Off Hits Hard Times." *Financial Times*, May 21, 1992, I12.
"Privatization Revenues Top HuF 26B." *MTI Econews*, May 27, 1992.
"Hungary: Privatization Proves to Be a Long, Hard Process." Inter Press Service, May 29, 1992.

"Government Gives Boost to Home Grown Capitalism." *Financial Times*, May 29, 1992.
"Hungarian Proposal Would Favor Domestic Over Foreign Investors." *BNA International Financial Daily*, June 4, 1992.
"Parliament Passes ESOP Law." *MTI Econews*, June 10, 1992.
"IBUSZ Papers for Compensation Coupons." *MTI Econews*, June 22, 1992.
"Hotel Sale Epitomizes Hungary's Sometimes Painful Privatization." *Mergers and Acquisitions Report* 5, no. 25 (June 22, 1992): 14.
"Companies to Retain State Interest." *MTI Econews*, June 24, 1992.
"Hungarian Parliament Approves New Agency to Manage Enterprises Retained by State." *BNA International Business Daily*, June 24, 1992.
"Hungary's Progress Criticised." *East European Markets*, June 26, 1992.
"Privatization of Large Banks." *The Hungarian Observer* (June 1992).
"Hungary: Income from Privatisation in First Five Months of 1992." *BBC Summary of World Broadcasts*, July 2, 1992.
"Survey of Privatisation in Eastern Europe." *Financial Times*, July 3, 1992.
"SPA Report on 1991." *MTI Econews*, July 9, 1992.
"Hungary Hotel Sale to Encourage Local Investors." *Reuters Business Report*, July 15, 1992.
"Public Placement of Danubius Hotel Shares." *MTI Econews*, July 16, 1992.
"Pick Salami Shares Aimed at Small Investors." *MTI Econews*, July 16, 1992.
"Marriott Buys Hungarian Hotel Stake." *Financial Times*, July 16, 1992.
"State Farms to be Privatized." *MTI Econews*, July 23, 1992.
"Employee and Management to Buyout Kanizsa Brewery." *MTI Econews*, July 24, 1992.
"Vagaries of Valuing Assets Pose Problems for Investors." *BNA International Business Daily*, July 27, 1992.
"Planned Disposal of Assets of State Property Agency." *BBC Summary of World Broadcasts*, Aug. 6, 1992.
"Foreign Investment in Privatization." *MTI Econews*, Aug. 11, 1992.
"Privatization Slows in Hungary." *MTI Econews*, Aug. 13, 1992.
"Revlon May Buy Caola." *MTI Econews*, Aug. 13, 1992.
"Future of SPA Head in Doubt." *MTI Econews*, Aug. 25, 1992.
"Permanent State Ownership in Commercial Banks." *MTI Econews*, Aug. 31, 1992.
"Prodded by New Law, Hungarian Banks Edge Away from State Control Toward Privatization." *EBRD Watch* 2, no. 32 (Aug. 31, 1992): 3.
"Bank Privatization Adviser to Be Chosen in October." *MTI Econews*, Sept. 8, 1992.
"SPA Official on Selection of Bank Privatization Adviser." *MTI Econews*, Sept. 8, 1992.
"Steady Growth in Personal Savings." *MTI Econews*, Sept. 17, 1992.
"Danubius Privatization—Further Details." *MTI Econews*, Sept. 22, 1992.
"EBRD Chasing Stakes in Hungary's Commercial Banks." *Reuters Money Report*, Sept. 29, 1992.
"Shortlist of Bank Privatization Consultants Published." *MTI Econews*, Sept. 30, 1992.
"Public Issue of Pick Salami Shares to Start October 27." *MTI Econews*, Sept. 30, 1992.
"Hungary Hunkers for Native Capital." *Budapest Week*, Sept. 3-9, 1992, 5.
"Hungary Makes Striking Switch to Mass Privatisation." *Financial Times*, Oct. 6, 1992.

"Marriott to Purchase Duna-Intercontinental." *MTI Econews*, Oct. 6, 1992.
"Finance Minister Says Privatisation at Standstill for Last 6 Months." *BBC Summary of World Broadcasts*, Oct. 6, 1992.
"State to Keep Ownership of 100 Enterprises." *MTI Econews*, Oct. 7, 1992.
"Hungary Readies Privatisation 'Credit Cards.'" *Reuters Library Report*, Oct. 12, 1992.
"Privatization Loan Coupon." *MTI Econews*, Oct. 14, 1992.
"BSE to Establish Clearing House." *Daily News* (Budapest), Oct. 2-8, 1992, 4.
Budapest Week, Oct. 8-14, 1992.
"Hungary Plans Further Public Shares Offering." *Financial Times*. Nov. 25, 1992.
"Business Briefs." *Daily News* (Budapest), Nov. 6-12, 1992, 4.
"SPA Strikes BZW Off Its Adviser List." *Daily News* (Budapest), Nov. 6-12, 1992, 4.
"Privatization Revenues Double." *Daily News* (Budapest), Nov. 6-12, 1992, 4.
"SPA Bounces British Bank in Beer Brawl." *Budapest Week*, Nov. 12-18, 1992, 2.
"Barclays and SPA Clarify Differences." *Daily News* (Budapest), Nov. 13-19, 1992, 4.
"New Privatization Forms." *Daily News* (Budapest), Nov. 13-19, 1992, 4.
"Danubius Share Issue Going Well." *MTI Econews*, Dec. 3, 1992.
"Reduced E-Credit Interest." *MTI Econews*, Dec. 11, 1992.
"Government Approves Plan to Speed Up Privatization." *MTI Econews*, Dec. 11, 1992.
"Hungary Acts to Speed Up Flagging Privatisation." *Reuter Library Report*, Dec. 11, 1992.
"Government on Economic Issues; SPA on Privatization Speed-Up." *MTI Econews*, Dec. 11, 1992.
"Alitalia Acquires 30 Pct of Hungarian Airline Malev." *AFX News*, Dec. 15, 1992.
"Privatization to Gain Momentum Next Year, Says Minister." *MTI Econews*, Dec. 22, 1992.
"Hungary Steps Up Sell-Offs." *MTI Econews*, Dec. 23, 1992.
"Concessions to Domestic Buyers." *MTI Econews*, Dec. 23, 1992.
"Privatization: Concessions to Domestic Buyers." *MTI Econews*, Dec. 23, 1992.
Budapest Sun, April 22, 1993, 1.
Budapest Post, April 22, 1993, 1.

Books

The World Bank, Hungary—Economic Developments and Reforms. 1984.
Berend, Ivan T. *The Hungarian Economic Reforms 1953-1988*. 1990.
Berend, Ivan T, and Gyorgy Ranki, eds. *The Hungarian Economy in the Twentieth Century*. 1985.
Bernat, Tivador, ed. *An Economic Geography of Hungary*. 1985.
Birenbaum, David, Leon E. Irish, and Karla W. Simon, eds. *Doing Business in Eastern Europe*. 1991.
Brada, Josef C., and Istvan Dobozi, eds. *Money, Incentives, and Efficiency in the Hungarian Economic Reform*. 1990.
Burant, Stephen R., ed. *Hungary—A Country Study*. 1990.
Fekete, Janos. *Back to the Realities*. 1982.
Hann, C. M., ed. *Market Economy and Civil Society in Hungary*. 1990.

Hare, Paul, Hugo Radice, and Nigel Swain, eds. *Hungary: A Decade of Economic Reform.*
Hieronymi, Otto. *Economic Policies for the New Hungary: Proposals for a Coherent Approach.* 1990.
Kornai, Janos. *The Road to a Free Economy.* 1990.
Kornai, Janos. *Vision and Reality, Market and State.* 1990.
Mero, Katalin. *The Role of the Stock Market and its Significance in the Economic Life of Capitalist Hungary (1864-1944).* Translated by Zsuzsanna Magulya.
Oppenheim, Klara, and Jenny Power. *Hungarian Business Law.* 1990.
Revesz, Gabor. *Perestroika in Eastern Europe—Hungary's Economic Transformation, 1945-1988.* 1990.
Richet, Xavier. *The Hungarian Model: Markets and Planning in a Socialist Economy.* 1989.

Scholarly Articles

"Legislation on Privatization and Acquisitions." *Doing Business With Eastern Europe*, Aug. 1, 1991.
Bartlett, David. "The Political Economy of Privatization: Property Reform and Democracy in Hungary." *East European Politics and Societies.* 6 (Winter 1992): 73, 106.
Bell, Joseph. "Privatization in Central and Eastern Europe." *PLI Course Handbook*, Sept. 25-26, 1991.
Blanchard, O. and R. Layard. "Economic Change in Poland," Discussion Paper No. 424, Center for Economic Policy Research, London, 1990.
Bod, Peter A. "Deregulation and Institution Building—Lessons From the Reform of the Hungarian Public Sector." *Jarbuch der Wirtschaft Osteuropas*, band 13, nr. 1 (1990).
Fallenbuchl, Zbigniew. "Polish Privatization Policy." *Comparative Economics Studies* 33 (1991): 65.
Matolcsy, Gyorgy. "Privatization: Hungary." *East European Economics* 30 (Fall 1991): 49, 52.
Ministry of Finance, "Preamble to the Act on Securities and the Stock Exchange." *Public Finance in Hungary.* 64 (1990): 3, 6-7.
Nagy, Bela Csikos. "Privatization in a Post-Communist Society—The Case of Hungary." *Hungarian Business Herald* 2 (1991): 36-37.
Newberry, David M. "Reform in Hungary—Sequencing and Privatisation." *European Economic Review* 35 (1991): 571, 575, 578-79.
Pechota, Vratislav. "Privatization and Foreign Investment in Czechoslovakia: The Legal Dimension." *Vand. J. Transnational Law* 24 (1991): 308.
Ringlien, Wayne. "Privatizing of Farming Activities in Hungary." *Finance and Development* (Dec. 1990).
Sajo, Andras. "Diffuse Rights in Search of an Agent: A Property Rights Analysis of the Firm in the Socialist Market Economy." *International Review of Law and Economics* 10 (1990): 41.
Schwartz, Gerd. "Privatization: Possible Lessons from the Hungarian Case." *World Development* 19 (1991): 1731, 1733.

Stark. "Privatization in Hungary: From Plan to Market or From Plan to Clan?" *East European Politics and Societies* 4 (Fall 1990): 351.

Sulkowski, Glick, and Richter. "Privatization in Hungary: The Art of the Possible." *International Financial Law Review* 10 (1991): 32.

Interviews

Katharine Ashton, January 1991.
Theodore Boone, January 1991.
Kalman Debreczeni, January 1991.
Dr. Gabor Faludi, January 1991.
Peter Fath, January 1991.
Katalin Gyurko, January 1991.
Prof. Lyman Johnson, January 1991.
Dr. Andras Kisfaludi, January 1991.
Tamas Lovas, January 1991.
Dr. Tibor Madari, January 1991.
Gabor Molnar, January 1991.
Dr. Zoltan Pacsi, January 1991.
Dr. Thomas Michael Revesz, January 1991.
Dr. Tamas Rusznak, January 1991.
Andras Simor, January 1991.
Norbert Streitman, January 1991.
Antonia Szabo, January 1991.
Dr. Mariann Vida, January 1991.
Dr. Zsuzsanna Zatik, January 1991.
Dr. Zoltan Pacsi, September 1991.
Antonia Szabo, September 1991.
Dr. Tibor Madari, January 1992.
Dr. Tibor Madari, September 1992.
Dr. Gabor Faludi, October 1992.
Dr. Andras Kisfaludi, October 1992.
Karoly Bognar, November 1992.
Dr. Jeno Czuczai, November 1992.
Dr. Viktoria Biro, February 1993.
Bela Jancso, February 1993.
Dr. Andras Kisfaludi, February 1993.
Antonia Szabo, February 1993.
Karoly Bognar, March 1993.
Dr. Maria Horvath, March 1993.
Zsuzsanna Kiss, March 1993.
Zsuzsanna Marton, March 1993.
Peter Padanyi, March 1993.
Antonia Szabo, March 1993.
Erich Wehr, March 1993.
Agnes Winkler, March 1993.
Peter Zelnick, March 1993.

Dr. Tibor Madari, April 1993.
Drs. Zsolt Sztrokay and Zsuzsanna Zatik, April 1993.

Miscellaneous

Law Decree 11 of 1986 on Liquidation Procedure; Decree No. 79/1988 on State Liquidation.
Bureau of Labor Statistics, U.S. Department of Labor. 1990.
Regulations of the Settlement Department of the Budapest Stock Exchange. 1990.
Charter of the Budapest Stock Exchange. 1990.
Rules of the Budapest Stock Exchange Regarding the Requirements of the Listing and Trading of Securities on the Stock Exchange. 1990.
Rules of the Budapest Stock Exchange Regarding the Types of Transactions That May Be Carried Out on the Stock Exchange and Trading on the Floor of the Stock Exchange. 1990.
State Property Agency. *First Privatization Program.* 1990.
State Property Agency. *The State Property Agency (SPA) Announces the First Privatization Program (FPP).* Sept. 1990.
State Property Agency. *Information About the Method of Reporting and Judgement of the Investor-Initiated Privatization of State-Owned Enterprises and State-Owned Shares of Associations.* 1991.
State Property Agency. *Information Booklet.* 1991.
State Property Agency. *Information Regarding the Second Privatization Program.* Mar. 1991.
State Property Agency. *Privatization and Foreign Investment In Hungary.* Mar. 1991.
State Property Agency. *Invitation for Tender on Managing Share and Stake Portfolios Owned by the State Property Agency.* April 1991.
Speech of Sir William Ryrie in Czechoslovakia, July 1991.
State Property Agency. *And Hungary?* (pamphlet). 1992.
Arms, Abbie. *Memorandum to Zoltan Pacsi*, a memorandum from SEC staff person to chairman of State Securities Supervisory Board suggesting changes in Hungarian securities laws. Oct. 29, 1992.
Debevoise and Plimpton. *Central and Eastern European Bulletin.* March 23, 1993.

Index

Accounting Act *see* Act XVIII of 1991
accounting/auditing 11, 22–23, 59, 60–61, 76, 84, 135, 167–69
Act VI of 1988 on Business Organizations 11–14, 15
Act VI of 1990 on Foundation and the Stock Exchange (Securities Act) 16–19, 20, 27, 119, 120, 167
Act VII of 1990 on the Foundation of the State Property Agency (Privatization Act) 19–20, 34
Act VIII of 1990 on the Protection of Property Entrusted to Enterprises of the State (State Property Protection Act) 21, 34
Act XIII of 1989 on the Conversion of Economic Organizations and Business Associations (Transformation Act) 15–16, 33, 47, 48
Act XVIII of 1991 on Accounting (Accounting Act) 22–23
Act XXIV of 1988 on Investments of Foreigners in Hungary 14–15, 167
Act XXV of 1991 on Partial Compensation for Damages Unlawfully Caused by the State to Properties Owned by Citizens in the Interest of Settling Ownership Relations (Compensation Act) 23–24, 155, 173
Act XLIV of 1992 on the Employees' Part-Ownership Program 102
Act IL of 1991 on Bankruptcy Procedures, Liquidation Procedures, and Final Settlement (Bankruptcy Act) 24
Act LIV of 1992 on the Sale, Utilization, and Protection of Assets Temporarily Owned by the State 80–81
Act LX of 1991 on the National Bank of Hungary (Hungarian National Bank Act) 25
Act LXIII of 1991 on Investment Funds (Investment Fund Act) 25–26
Act LXIX of 1991 on Banks and Banking Activities (Banking Act) 26, 86
Act LXXIV of 1990 on the Privatization of State-Owned Companies Engaged in Retail Trade, Catering, and Consumer Services 21–22
Active Privatization Program, SPA's, Phase I *see* First Privatization Program
Active Privatization Program, SPA's, Phase II *see* Second Privatization Program
Active Privatization Program,

SPA's, Phase III *see* Third Privatization Program
Active Privatization Program, SPA's, Phase IV *see* Fourth Privatization Program
Africa 5, 113
Agreement on Trading in Securities 117
agricultural holdings 2, 23-24, 39, 81, 82, 83, 158, 166, 173; privatization of state-owned farms 5, 93-95
AGROINVEST Rt. 102
Alitalia 108
Alliance of Free Democrats (SDS) 33, 40, 54, 133, 145
Alliance of Young Democrats (Fidesz) 33, 40, 54, 133, 145
Antall, Jozsef 45, 153, 166
Antall, Laszlo 39, 40, 136
Apisz transaction 32-33, 47, 49
Arent Fox 108
articles of association 11, 12, 15
Asia 5
asset leasing 20, 41, 63-64, 67, 72, 80-81, 154, 158
asset management schemes 41, 73, 142
auctions 57, 66-69, 71, 142, 173; Hungarians-only 20-21, 66-67, 71, 103, 137, 161
Austria 5, 17, 32, 37, 42, 43, 50, 61, 68, 113-14, 146, 163; Creditanstalt 26, 43, 45, 128; Creditanstalt CA Investment Fund 26, 127-28; Girozentrale 42-45, 99; Vienna Stock Exchange 44, 112, 113; Wiener Allianz Versicherungs-Aktien-gesellschaft 88
Autoklinik Group 91
AVB *see* General Bank of Venture Financing

banking 67, 82-93; bailout 92-93; collapse 88, 89-92, 113; commercial 25, 26, 28, 67-68, 82-88, 92, 100, 117, 147, 167; malpractice 89-90; privatization 25, 71, 85-88; reform 8, 25-26, 67, 82-85, 117, 167-68; savings bank 82-84, 158
Banking Act *see* Act LXIX of 1991
Bankruptcy Act *see* Act IL of 1991
bankruptcy law 24, 79
Barclays de Zoete Wedd (BZW) 169-70
Barque Indosuez (France) 88
Bartlett, David 5, 29, 168
Bethlem, Istvan 45
Bod, Peter A. 86, 115, 136, 150, 167
Bokros, Lajos 5, 59, 60
Bolivia 40
bonds 16, 33, 41, 92, 113, 115-17, 118, 119, 127
book entry 129
Borsodtavho 159
brokerage firms 17-18, 27, 135
Bt. *see* limited partnerships
Budapest Bank 43, 83, 84, 116-17, 118
Budapest Chamber of Commerce and Industry 113
Budapest Commodities Exchange (BCE) 128
Budapest Commodity and Capital Exchange 111-14; Exchange Council 111
Budapest Stock Exchange (BSE) 8, 10, 17, 18, 26, 27, 28, 44, 46, 57, 59, 60, 62, 89, 109, 110-31, 150-52, 155, 167, 176; activities 122-31; establishment and organization 117-22; history 111-17; index 130-31; membership 18, 118, 120-21; requirements, general 110, 124, 131; Rules of the Budapest Stock Exchange 121, 123; Securities Trading Committee 118, 121-22, 129; Stock Exchange Arbitration Court 122; Stock Exchange Council 27, 117-18; 119, 121,

122, 123, 129; Stock Exchange Ethics Committee 122; Stock Exchange Supervisory Committee 121
Budapest Wheat Hall 111
budget deficit 39–40, 156–57
Bulgaria 6, 133–34
business associations 11–12
business organizations, formally recognizing 11–14
Byelorus 6

Canada 146
Caola 75–76, 140
capital, defined 2
capital market regulation, defined 2
capital markets, defined 2
Castellum Ltd. 108
Central Statistical Office 134
certificates of deposit 119
Christian Democrats 40
Co-Nexus 74
Colgate-Palmolive Company 75–76
collieries 83, 147
COMECON 38, 42, 50, 164
Communists 51, 63, 114, 152, 165
compensation 23–24, 39–40, 94, 148, 172–73
Compensation Act *see* Act XXV of 1991
compensation coupons 23–24, 46, 104, 105, 127, 128, 155, 158
competitive bidding 20, 22, 52, 57, 70–72, 138, 140, 142–43, 147, 149, 164, 169, 172, 178; special favorable consideration for domestic bidders 99, 103–6
Constitutional Court 23, 40, 54
construction 61, 65–66, 73
consultancy services, postprivatization 77
consumer services businesses 21–22
cooperatives 15, 37, 83, 94
credit 82, 147–49; allocation 83; consolidation 88; easing for domestic investors 99–101, 104, 141, 149, 161, 163; Existence Credit (E-Credit) 100–2; paucity of 25, 67–68, 85, 97, 135, 147–48, 162, 163, 167, 177; Privatization Credit 100; *see also* loans
credit cards 41, 101
Creditanstalt *see* Austria
Croatia 6, 168
Csemege, Julius Meinl 131
Csepi, Lajos 45–46, 59, 63, 75, 87, 101, 104, 107, 137, 144, 150–51, 169, 170
Csuhaj, Imre 88
Csurka, Isran 165–66
currency control 83; exchange 61, 74, 83, 179
Czechoslovakia/Czech Republic 6, 41, 53, 105, 108, 148, 178

Daiwa Securities 74
Danubius Hotal & Spa Co. 55, 60–62, 104, 109–10, 126, 131, 140, 151, 155, 166; Budapest Hilton 60; Gellert Hotel 104, 166
Deak, Imre 62
debt-for-equity swap 40, 42, 104, 155
debt securities 16, 115, 127, 128, 130, 131, 162
decentralization of privatization 152–54, 175
Dekany, Sandor 140
Deloitte and Touche 89
democratic government/reforms 3, 5, 29, 33, 38, 45, 51, 66, 69, 133, 150, 166
deposit insurance 92
domestic participation 96–109, 146–47, 149, 177; difficulties of 15, 25, 105, 127–28, 134, 144, 147; education to encourage 99, 123; encouraging/ensuring 22, 24, 96, 106; facilitating 26, 53,

66, 70, 80–81; incentives, generally 41, 98, 99–106, 141, 147–48, 177; lack of demand 97–99, 105, 119, 141, 161–63, 177; legal guarantees 19; maximizing demand 15, 137, 141, 147–48, 177, 179
domestic savings 68–69, 92, 96, 97–98, 127, 147–48, 149, 161–62
domestic share set-asides 41, 103, 161
Dorog Coal Mines 147

Economic Cabinet 146, 153, 165; Ownership and Privatization Committee 153
Electrolux 76
Employee Share Ownership Plans (ESOPS) 40, 57, 102
employees, preferences for 20, 71, 77, 99, 101–3, 104, 108, 155
enterprise councils 33, 37
entrepreneurial activity 8, 38, 68, 69, 85, 91, 100–1, 106, 108, 134, 139, 147, 172, 178
environmentalism 173–74
equipment, capital 2, 3, 15, 38, 39
Estonia 6
Eurobond 41
European Bank for Reconstruction and Development 77, 87
European Community's Insider-Dealing Directive 18

Farbod, Lotfi 129
Farkas, Karoly 88
Fekete, Janos 115
Fidesz *see* Alliance of Young Democrats
Finance Ministry *see* Ministry of Finance
First Asset Manager Tender 72–74
First Privatization Program (FPP) 54–62, 63, 67, 69, 71, 75, 138, 140, 150–51, 152

foreign aid 9, 34, 156, 165, 176
foreign investment 25, 80, 86, 93, 96, 105, 107, 124, 135, 144–48, 149, 160–61, 166, 168, 172, 173, 177; benefits of 179; competition for 108; demand 45, 69, 96, 105–8; direct 107–8, 146, 168, 178; discouraging 39, 72, 139; encouraging 13–14, 32, 33, 53, 56–58, 77, 96, 137; fear of domination 20, 54, 61, 66, 99, 103, 135, 141, 143, 144–45, 146, 149, 160–61, 168, 177, 179; incentives 179; ineligibility 22, 95; lack of demand 41, 161, 163; legal guarantees 11, 19, 176, 179; maximizing demand 179; necessity/desirability 145–46, 162, 177, 179; political problems with 160–61; protection of 11, 14–15; reliance on 48, 70; requiring 26; restrictions on 14, 85, 93, 137; shift away from 104; targeting 17, 69
foreign/national debt 32, 39–40, 133, 136, 148, 156–57
Fotex 125, 126, 127
Fourth Privatization Program 65–66, 71
France 5; Paris stock market 113
Fransaholding 88
fraud 2, 89–90, 118
Friend, Adrian 58
Fry, Fiona Somerset and Plantagenet 4
Fund Transfer Policy, SPA 87
futures 127, 128

Ganz Electric Motors 48
Ganz Engineering 48
General Bank of Venture Financing (AVB) 90–92
General Electric 42–43; *see also* Varga, George
general partnerships (Kkt.) 11, 12
Germany 58, 68, 90–92, 100, 112, 147; German AG 13; German

Index

GmbH 13; GFT Baumler, AG 151; Konrad Adenauer Foundation 96, 138; Westdeutsche-landesbank (WDLB) 90–92; *see also* Springer, Axel
giveaways 39–40, 53, 70, 97–99, 105, 133, 136, 138, 143, 148–49, 177–78, 179
Graboplast 43–44
Grenfell, Morgan 102
gross domestic product (GDP) 5, 47, 48, 134, 136
gross national product (GNP) 106
Grosz, Karoly 32
Gyomaendrod Savings Bank (GSB) 90

HAFE 79
hard assets 98, 119, 161, 162
Hardy, Ilona 117, 118, 123, 129
Hartai, Ervin 69
holding companies 33, 39, 49, 80; hollow parents 80; shell companies 31, 48, 63, 80
Hollo, George 97, 144
hostile takeovers 71–72
hotels 14, 56, 61, 140; *see also* Danubius Hotel & Spa Co.; HungarHotels; Pannonia Hotel and Catering
HungarHotels 31–33, 47, 50, 55, 140, 141
Hungarian Association of Entrepreneurs 147
Hungarian Chamber of Commerce 117
Hungarian Civil Code 11, 16, 129
Hungarian Commercial and Credit Bank 83, 84
Hungarian Commercial Bank 46
Hungarian Company or Corporation Law *see* 1988 Act on Associations
Hungarian Credit Bank 42, 83, 84, 87, 140
Hungarian Democratic Forum (MDF) 33, 34, 45, 53, 61, 70, 80, 93, 99, 103, 106, 115, 133, 135, 136, 137, 139, 145, 152, 153, 165, 169
Hungarian Foreign Trade Bank 84
Hungarian Governor's Council 111
Hungarian Investment and Development Company (MBF Rt.) 92
Hungarian Kulturbank 88
Hungarian National Bank (HNB) 25, 26, 28, 32, 41, 57, 61–62, 67, 68, 83–87, 90, 92, 97, 100, 101, 115, 117, 121, 136, 144, 162
Hungarian National Bank Act *see* Act LX of 1991
Hungarian Parliament 10–28, 30, 44, 66, 75, 88, 90, 102, 103, 108, 115, 152, 158, 171–74
Hungarian Post Office 115–16, 158
Hungarian Socialist Workers Party (HSWP) 8, 11, 29, 32, 33, 34, 36–37, 50, 53, 83, 98, 114, 115, 137, 146, 152; *see also* Nemeth, Miklos regime
Hungarian State Oil and Gas Trust 115, 116
Hunglet 48

IBUSZ 44–47, 54, 55, 59, 60, 62, 69, 110, 125, 126, 131, 137, 141, 143, 150, 169; *see also* Szemenkar, Ericka
Independent Smallholders party 39, 53, 93, 135, 145
India 5
industry 33, 56, 57, 64, 83–84, 113, 136–37, 153, 173
inflation 40, 68, 74, 114, 115, 119, 130, 131, 157, 178
infrastructure 8, 19, 27, 65, 156, 163, 167–68, 176, 179
insider-dealing 18
institutional ownership 40
intellectual property laws 167
interest rates 8, 25, 41, 61, 68, 69,

90, 93, 98, 100–1, 106, 115, 119, 147, 149
International Monetary Fund (IMF) 156, 165
Investment Fund Act *see* Act LXIII of 1991
investment funds 9, 25, 66, 98, 100, 104, 127–28, 131, 147, 162, 177; closed-end funds 25–26, 127, 131; CA (Creditanstalt) Investment Fund 127–28; open-end funds 25–26
investor-initiated privatizations 70–72, 142, 166
Iran 129
Iraq 168

Jarai, Zsigmond 32, 116–17, 118
joint stock companies 33, 48, 63, 70, 141, 151
joint ventures (Kv.) 11, 12–13, 30–31, 48, 49–50, 63, 80, 134
judicial review of company resolutions 11

Kadar, Janos: regime 7, 29, 114–15; *see also* New Economic Mechanism
Kanizsa Brewery 101–3, 147
Keskeny, Ildiko 68
Kft. *see* limited liability companies
Kkt. *see* general partnership; unlimited partnerships
Kobanya Brewery 169
Konzumbank 88
Kornai, Janos 136
Kupa, Mihaly 165
Kuwait 168
Kv. *see* joint ventures

Latvia 6
Law on Enterprise Councils 32
Lehel Manufacturing 76
Lengyel, Laszlo 166

Liberal party 111
limited liability companies (Kft.) 12, 13, 15, 22
limited partnerships (Bt.) 11, 12
liquidation 24, 49, 64, 89, 171; privatization through 78–80, 108, 171; spontaneous 79
listed securities 124–27, 131
Lithuania 6
loans, bank 67–68, 69, 83–84, 85, 90, 92–93; coupons 101; high-risk 92; loan-for-bond exchange 92–93; nonperforming 22, 84, 92, 167; preferential for domestic investors 41, 57, 99–101, 104; subsidized 61–62; unsecured/uncollectable 84, 89–90, 91, 92; *see also* credit
local councils 37, 60–61, 102, 103, 104, 155

Malev Airlines 108, 159
managers: fiduciary duties 21, 25; preferences for 20, 71, 77, 81, 99, 101–3; self-serving practices 48–51
Martonyi, Dr. Janos 98, 146
MATAV 109
Matolcsy, Gyorgy 38, 45, 134
MDF *see* Hungarian Democratic Forum
Medicor 48
Merrill Lynch 31–32, 58, 136, 163; *see also* Friend, Adrian
Mexico 40
middle class, creation of 38, 66, 69, 97, 98, 133, 150, 162, 177
Minister Without Portfolio for Privatization 87
Ministry of Finance 14, 76, 79, 87, 118, 142, 153, 157, 165
Monor farm 94–95
Muszi 125, 126, 129
mutual intercorporate ownership 40–41

Index

Nemeth, Miklos: regime 29, 34, 35, 50; *see also* Hungarian Socialist Workers Party
New Economic Mechanism 7, 29, 32; *see also* Kadar, Janos regime
Newberry, David 134
1984 Act on Enterprise Councils 33, 47
1988 Act on Associations (Hungarian Company or Corporation Law) 33, 47, 48, 87, 167
non-profit associations 12, 40

Office of State Commissioner on Privatization 34
options 127, 128
organic development of market economies 2–6, 136
over-the-counter market 110, 111, 118, 119, 124, 131, 151
ownership, lack of clarity 23–24, 36–37, 49, 67, 94–95, 107, 154–56, 166, 173, 176–77
Ozd 48

Pacsi, Dr. Zoltan 27
Palotas, Janos 147
Pannonia Hotel and Catering 55, 140
pension funds 57
performance figures 22–23, 36, 37, 134, 135, 142, 164
Pest Chamber of Commerce 111
Pest Lloyd Company 111
Petrenko, Janos 48
Pharmatrade 147
Pick Szeged 103, 104, 109, 110, 126, 131, 151, 155, 159
Poland 6, 39, 41, 53, 108, 148, 178; voucher system 97–99
Policy Guidelines on Assets 101
Posta Bank 88
Price Waterhouse 45
private placements 17, 25, 28, 98, 109, 110, 150–52, 179; pre-SPA 42–44
Privatization Act *see* Act VII of 1990
privatization advisers 57–60, 76–77, 88, 153, 154, 169–70
Privatization Research Institute 96–97, 138
professional class 98
prospectuses 17–18, 27, 170; pre-SPA 42, 44–47
public companies limited by shares (Rt.) 12, 13, 15, 118, 147
public offerings 17–19, 23, 25, 27–28, 54, 57, 62, 109, 141, 150–52, 155, 168, 169, 179
public utilities 71

Quintus 31, 141

Rajcsanyi, Peter 59, 62
recession 153
reform legislation 6–8, 10–26, 34, 95, 118, 176; conflict of law therein 171–74, 179
Reorg PLC 79; *see also* State Liquidation Agency
residual state ownership 157–59, 179
restaurants 21–22, 25, 37, 58, 60, 66–69, 71, 103, 137, 161, 163
risk reserves 86
Romania 6
Rt. *see* public companies limited by shares
Russia 6

Sachs, Jeffrey 39
Schlumberger Industries 48
Schwartz, Gerd 154
SDS *see* Alliance of Free Democrats
Seagrams 103
Second Asset Manager Tender 74

Second Privatization Program (SPP) 63–65, 71, 138
Secretariat for Trade in Securities 117
Securities Act *see* Act VI of 1990
securities markets, development of 110–31
self-privatization 142, 153, 154, 166
Self-Privatization Program 76–78, 153, 154
Serbia 168–69
Simor, Andras 135, 146
SIMSET 108
Slovakia *see* Czechoslovakia
Slovenia 6
small shops 21–22, 25, 37, 58, 66–69, 71, 103, 137, 161, 163
Social Democrats 40
Social Security System 46, 91, 138, 157, 158
Socialism/Socialist countries (former) 3–8, 39, 40, 43, 115, 119, 148, 149, 175
South America 5
Soviet Union 3, 38, 50, 79, 94, 164
spontaneous privatizations 21, 30–32, 33, 34, 59, 63, 80, 140; pre-SPA 30–34, 42, 47–52, 63, 75, 133, 137, 141, 152; pre-SPA abuses 48–51; SPA sponsored 47, 51–52, 71, 74–78, 176
Springer, Axel 141
State Asset Holding Company (SAHC) 108, 142, 158
State Audit Office 20, 35
State Banking Supervisor (SBS) 85, 87, 89–91
State Development Bank 115–16
State Development Institute 121
State Liquidation Agency (SLA) 79; *see also* Reorg PLC
State Privatization Institute 142
State Property agency (SPA) 5, 10, 15, 17, 20, 21, 24, 26–27, 28, 29–36, 42, 44–69, 70–81, 87, 91, 94, 96–110, 128, 131, 137–46, 150–58, 164–66, 169–72, 176; creation 20, 30–32, 42; staffing 30, 35–36, 60, 71, 141–42, 169, 171, 176; structure 20, 35; tasks 20, 34
State Property Protection Act *see* Act VIII of 1990
State Securities Supervision Board (SSSB) 89, 167
State Securities Supervisor (SSS) 8, 10, 17, 19, 26, 27–28, 123, 176
Styl 125, 126, 151
subsidiaries 49, 74, 80
subsidies 147–49, 159; agricultural 93–94; subsidized sales 105, 147–49, 178
supervisory boards/commissioners 11, 13, 89, 90
Szabo, Karoly 59, 158, 169
Szabo, Tamas 87, 105
Szakolszay, Gyorgy 96, 138
Szemenkar, Ericka 44

Tardos, Dr. Morton 133
Tauber, Jakab 112
tax incentives 14–15, 61, 104, 178
technology 57, 134, 146, 167
telecommunications 158, 171–72; Hungarian Telecommunications Company (HTC) 172; Ministry of Transport, Telecommunications, and Water Management 172; Telecommunications Act 171–72; telephones 115–16, 172
temporal parameters 7
Third Privatization Program 65–66, 71, 73
Tompe, Istvan 45–46, 137, 139, 143, 169
tourism 56, 68, 83
trade associations 12
traded securities 124–27, 131
transactions 2, 3; clearance and settlement of 128–29
Transformation Act *see* Act XIII of 1989

transformation of enterprises 15–16, 58, 61
Tungsram 42–43, 47, 49
Turi, Jozsef 68

Ukraine 6
unemployment 78, 94, 153, 156, 159
United Kingdom 3, 4, 5, 7, 34, 60, 134, 151; Know How Fund for Hungary 34; London Stock Exchange 60, 113, 118; Nomura Investment Bank 60–62
United Nations 168–69
United States 3, 4, 5, 7, 9, 13, 43, 113, 120, 128, 159; CFTC 128; EIA Acquisition 75; NYSE 114; SEC 120, 128; Support for East European Democracy Act (SEED) 9; United States Agency for International Development (USAID) 94
universities 138
unlimited partnerships (Kkt.) 11, 12

valuations: book 38, 47, 77, 78, 107, 139, 148; devaluation 140; difficulty in 22, 36, 37–38, 50, 59, 135, 148, 163–64, 166, 167, 168, 172; facilitating 23; "fair" pricing 53–54, 70, 135, 177; lack of 30, 41; market 36, 45, 52, 58, 62, 70–71, 74, 76, 139, 141–43, 148, 153, 163–65, 166, 168, 169, 172, 177; overvaluation 38, 67, 75, 137, 148, 164–65, 166; self-serving 48–49; undervaluation 31, 38, 49, 141, 164
Varga, George 43
Vegytek 75, 140
Veszprem Catering Company 68
Videoton 140
Vitali 140
vouchers 97–99, 105

warrants 128
wealth, dispersal of 40, 41, 98, 115, 162, 177
wine regions 65–66
worker councils 15–20
World Bank 9, 34, 118

Ybl Bank 89–90

Zalakeramia 125, 126, 151
Zwack, Peter 103
Zwack, Unicum 103, 131